NRA

National Rivers Authority

CONTAMINANTS ENTERING THE SEA

A REPORT ON CONTAMINANT LOADS ENTERING THE SEAS AROUND ENGLAND AND WALES FOR THE YEARS 1990 - 1993

Report of the National Rivers Authority

May 1995

Water Quality Series No. 24

LONDON: HMSO

CONTENTS

FIGURES AND TABLES

EXECUTIVE SUMMARY

The UK is required through various international agreements and commitments to reduce the quantities of certain hazardous substances and nutrients entering the sea from land-based sources. These substances have been selected for priority control because of their toxicity, persistence in the environment and potential to accumulate in marine life. Nutrients such as nitrogen and phosphorus have also been included in these commitments because of concerns about their potential under certain conditions to cause unnatural and excessive algal growth in estuaries and the sea. This report presents the progress that has been made in reducing the quantities of these contaminants entering the sea around the England and Wales coastline up to 1993.

The UK is a Contracting Party of the 1974 Paris Convention, which is aimed at preventing and reducing marine pollution from land-based sources (and which will be replaced by the 1992 OSPAR Convention when that has been ratified by all concerned). The UK has also joined in the commitments in the North Sea Declarations which, among other things, are aimed at achieving significant reductions (of 50% or more) of the inputs of specified contaminants to the North Sea between 1985 and 1995. In England and Wales, the National Rivers Authority (NRA) is responsible for monitoring the inputs of these listed contaminants. It does this by measuring the loads of substances from direct industrial and wastewater discharges into estuaries and coastal waters, and by assessing the loads of substances entering the sea via rivers at the tidal limit.

The monitoring programme has provided some important information on patterns and trends in inputs of contaminants to the sea over the last few years. It is clear that the relative contributions from different sources varies between different contaminants according to their pattern of use. For example, the main source of the metals mercury, cadmium, arsenic and chromium is direct industrial discharges to coastal waters, whereas a significant part of the overall input of nitrogen, copper, lead, zinc and nickel is carried via rivers, much of which originates from diffuse sources within river catchments.

The levels of contaminant inputs monitored over the period 1990 to 1993 varied around the coastline; the North Sea and Irish Sea generally receiving the highest loads of many substances. Loads of copper, zinc, lead, nitrogen and phosphate were highest into the North Sea, whereas cadmium and mercury inputs were greatest into the Irish Sea.

Although inputs have been assessed using consistent methods since 1990, the data available for 1985 are limited, making it difficult in some cases to define a reliable baseline. However, where it has been possible to establish a baseline estimate for 1985, some clear trends have emerged. For the North Sea there have been substantial reductions in loads of mercury and cadmium and some reductions in the loads of copper, lead, chromium and gamma HCH. Considering the overall inputs to the sea around the whole England and Wales coastline, there have been significant reductions in the loads of many of the priority metal contaminants. The exception to this is zinc, for which inputs have not significantly decreased since 1985.

The NRA is continuing to address the need for pollution prevention and control of priority contaminants where this is needed to meet the required input reductions. Relative inputs of different contaminants have been quantified and prioritised as part of an overall strategy to reduce inputs from point sources and deal with, as far as possible, inputs from diffuse sources. This is being done within the context of the UK and European regulatory framework, drawing upon a range of different available mechanisms for prevention and control.

Chapter 1   INTRODUCTION

The UK is required to control the discharge of hazardous substances to water by various international agreements and commitments. For example:

- **the Paris Convention;**
- **the North Sea Conference Declarations;**
- **European Community Directives.**

Figure 1.1 shows how these agreements interrelate. Such agreements and commitments typically control the input to the aquatic environment of certain substances which are specifically listed within the Convention, Declaration or Directive.

In order to meet these obligations, the Government requires inputs to estuaries and coastal waters to be monitored for the listed substances. In England and Wales, the National Rivers Authority (NRA) is the organisation responsible for this monitoring.

Since 1990 data arising from these surveys have been collated centrally within the NRA and then passed onto the DoE. So far the results from the surveys undertaken in England and Wales have not been published separately though they are available along with the rest of the NRA's analytical results on its public registers. Reports on the results from Europe as a whole are periodically produced by organisations such as the Paris Commission (PARCOM).

This report will present information gathered on an annual basis from surveys for Paris Convention and North Sea Conference Declaration (NSCD) purposes.

In addition to presenting survey data gathered during the 1990-1993 period, this report also aims to "set the scene" with regard to:

- **why this information is needed;**
- **the nature of the contaminants;**
- **their types and sources;**
- **how loads are calculated;**
- **what load reductions have already been achieved.**

The report does not differentiate between the various sources of contaminants in a river catchment to any great extent. However, this is addressed by the NRA in other ways; most specifically in the Catchment Management Plans it publishes and more generally by its routine day by day pollution prevention work. The results of the 1990-1993 surveys are then compared with 1985, where possible, and finally future actions and developments are discussed.

The input of contaminants to the sea does not necessarily result in pollution. A contaminant is a substance which would not normally be present in the sea and is only there as a result of some activity by man. If, however, a contaminant is present to such an extent that there is a hazard to human health, or an adverse effect on marine life, then this is pollution.

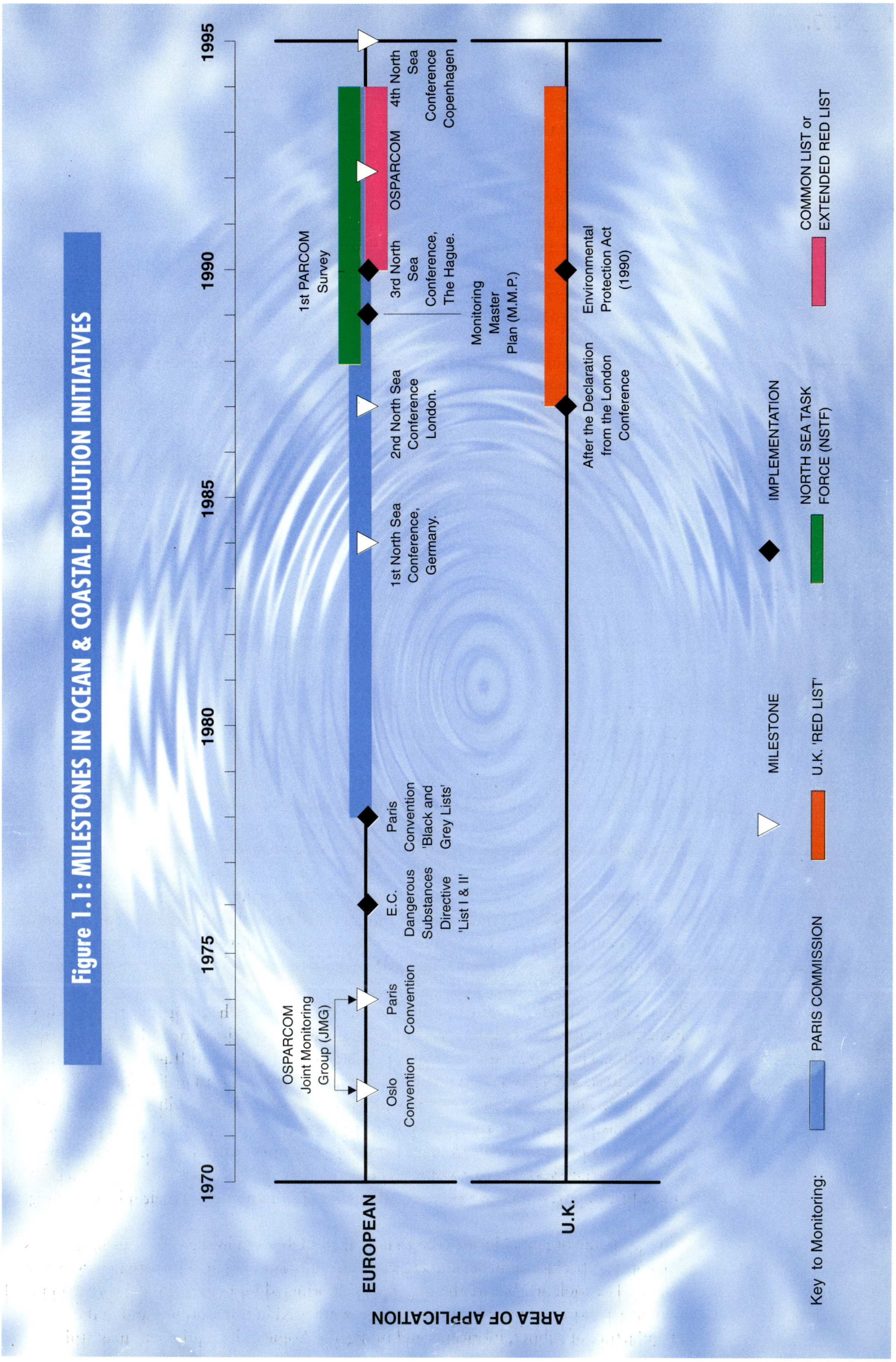

Figure 1.1: MILESTONES IN OCEAN & COASTAL POLLUTION INITIATIVES

Chapter 2   WHAT ARE THE CONTAMINANTS?

The chemical contaminants of interest for PARCOM and NSCD purposes can be split into a number of types or categories that have similar basic characteristics:

- inorganic metals;
- organometals;
- organophosphorus pesticides;
- organochlorine pesticides;
- other pesticides;
- other organochlorine compounds;
- nutrients.

The hazardous substances above have been selected for priority control and reduction on the basis of their toxicity, persistence and bioaccumulation within the aquatic environment. Nutrients have been included because of concern about eutrophication (enrichment of natural waters).

Substances included in the priority lists have a wide range of industrial, agricultural and domestic uses (see Table 2.1) from which they may find their way into rivers, estuaries and coastal waters. They may arise from single, 'discrete' points such as factory waste-pipes discharging into a river and/or 'diffuse' sources such as run-off from land or in rainfall. Occasionally inputs of chemical substances may occur through illegal use or disposal.

INORGANIC METALS
(eg. chromium)

Metals are used in a variety of industries from the production of alloys and metal plating to the manufacture of glass, wood preservatives, paints, agrochemicals and textiles. They are also used as chemical catalysts and in the production of dyes and pharmaceutical products.

Diffuse inputs to the aquatic environment (eg. from atmospheric deposition or surface run-off) may, for some metals, be greater than direct inputs from industry. Even so, direct inputs still represent considerable loads in terms of tonnage. Other sources include mining activities and landfill operations. Inputs of metals also arise from the use in the home of products containing these metals (eg. ointments, cleaners etc.) which then find their way into domestic sewage. Some metals (eg. zinc) are found at high concentrations in water due to natural sources (at low pH).

ORGANOMETALS
(eg. organotin compounds)

Organotins are used primarily in boat hull antifouling paints - although the use of tributyltin (TBT) for this purpose is now restricted in the UK - and, to a lesser extent, as stabilisers in plastics. TBT is also used as a wood preservative. The use of organotins contributes to both point and diffuse source pollution.

PESTICIDES (eg. DDT)

Most organochlorine and organophosphorus pesticides on the priority lists are, or have been, insecticides used in the UK. The others (ie. atrazine, simazine and trifluralin) are herbicides. The application of pesticides to land can give rise to diffuse inputs; eg. from spray drift, leaching and surface run-off and farm wastes. Additionally, inputs can arise from point sources. In some cases the diffuse inputs of pesticides will exceed the input from industry and domestic sources. In other cases the use of pesticides will provide a significant point source contribution (eg. pentachlorophenol and lindane from timber treatment processes). These point sources can arise from the processing of products (including imports) such as fleeces or other animal skins as part of their initial cleaning.

OTHER ORGANOCHLORINE COMPOUNDS
(eg. chloroform)

Non-pesticide organochlorine compounds are widely used in the chemicals industry as solvents (eg. trichloroethane) and chemical intermediates (eg. dichloroethane). Some are also used as coolants. Before the use of polychlorinated biphenyls (PCBs) was abandoned they had a variety of uses; for instance they were used in transformers and in the manufacture of rubber, lubricants and fumigants. Some older equipment may still incorporate PCBs, although they are no longer used in the manufacture of new equipment. Also, they can be formed as by-products in the manufacture of organic chemicals.

Organochlorine compounds are widely used in industry and, therefore, industrial effluents are likely to be a significant source of input. However, diffuse input and atmospheric deposition of organochlorine compounds are an important route for a number of substances.

NUTRIENTS (eg. nitrogen)

The main use of nitrate and phosphate in the UK, is as fertilisers in the agricultural industry. However, significant quantities of phosphates are also used as 'builders' in detergents to improve cleaning efficiency. Domestic sources are also important.

The principal routes by which nutrients reach rivers in agricultural catchments are diffuse: eg. the direct run-off of and leaching of fertiliser from farmland and afforested upland areas; and indirect run-off from intensive farming practices such as muck spreading. Atmospheric inputs, soil erosion, the release of nutrients from sediments and organic plant wastes also represent diffuse sources of nutrients.

In urban catchments point source nutrient inputs are more important. These are derived principally from sewage treatment works effluents and industrial wastes. A major source of phosphate is via the use of detergents. Up to 50% of the total phosphorus content of sewage maybe detergent-derived. The proportion of detergent-derived phosphate in rivers has been estimated at up to 25%.

Table 2.1 Sources and uses of substances included in UK priority lists for PARCOM and NSCD purposes

SUBSTANCE	SOURCE	USES
METALS		
Arsenic	Point and diffuse	Wood preservation, manufacture of glass, alloys, medicines and semi-conductors, by-product of smelting industry.
Copper	Point and diffuse	Metal plating industry, agriculture, manufacture of alloys, copper wire & piping, textile dyeing, glass & ceramics, catalyst in vinyl chloride production, manufacture of wood preservatives, rayon, paint pigments, active ingredient in marine antifouling paint.
Zinc	Point and diffuse	Manufacture of alloys, electroplating and galvanising, manufacture of rayon textiles, production of paper, fungicides, rubber, paint, ceramics, glass, reprographic materials, hygiene products.
Cadmium	Point and diffuse	Manufacture of pigments & stabilisers, batteries, cement, agrochemicals, alloys, solders, photoelectric cells, electrodes, electroplating, photographic processes, deoxidiser.
Mercury	Point and diffuse	Manufacture of batteries, agrochemicals, pharmaceuticals, mirrors, thermometers, barometers, chlor-alkali production, catalysts, dentistry.
Lead	Point and diffuse	Manufacture of batteries, production of anti-knock agents for petrol, cable covering, solder, pigments, type-metal, building materials, radiation shields, cement manufacture, shot for shooting and fishing.
Nickel	Point and diffuse	Metal plating industry, iron and steel production.
Chromium	Point and diffuse	Metal plating industry, iron and steel production, pigment production, textile colouring, lithographic and photographic applications, glass manufacture, ceramics production, leather tanning.
ORGANOPHOSPHORUS PESTICIDES		
Parathion	Mainly diffuse	Agriculture (never approved in the UK)
Parathion-methyl	Mainly diffuse	Agriculture (never approved in the UK)

SUBSTANCE	SOURCE	USES
Azinphos-methyl	Mainly diffuse	Agriculture (withdrawn in the 1990's in the UK)
Azinphos-ethyl	Mainly diffuse	Agriculture (withdrawn in the 1980's in the UK)
Fenthion	Mainly diffuse	Agriculture and veterinary (never approved in the UK)
Fenitrothion	Mainly diffuse	Agriculture, public health and domestic
Dichlorvos	Mainly diffuse	Agriculture, public health and domestic
Malathion	Mainly diffuse	Agriculture and domestic
ORGANOCHLORINE COMPOUNDS		
Carbon tetrachloride	Point and diffuse	Petroleum refining/coal processing, halogenation of non-aromatics, fabric mills, paper and pulp mills, manufacture of plastics, synthetic rubber, pharmaceuticals, flavourings, perfumes, cosmetics, industrial organic chemicals, electronic components.
Hexachlorobenzene	Mainly point	Biocide and chemical synthesis
Hexachlorobutadiene	Point	Solvent, refrigeration systems, hydraulic systems, transformer oil, dielectric fluid.
Polychlorinated biphenyls (PCBs)	Point and diffuse	Heat exchange agent, dielectric fluid, lubricant.
1,2-Dichloroethane	Point	Intermediate in the production of chlorinated hydrocarbons eg. vinyl chloride, 1,1,1-trichloroethane, trichloroethylene and tetrachloroethylene, solvent.
1,2,3-Trichlorobenzene	Point	Solvent and chemical synthesis
Tetrachloroethylene	Point and diffuse	Solvent and chemical synthesis
Trichloroethylene	Point and diffuse	Solvent and chemical synthesis
Trichloroethane	Point and diffuse	Solvent
Chloroform	Point and diffuse	Solvent and chemical synthesis
ORGANOCHLORINE PESTICIDES		
Aldrin	Mainly diffuse	Agriculture (use prohibited 1989)
Endrin	Mainly diffuse	Agriculture (use prohibited 1984)
Dieldrin	Point and diffuse	Agriculture, industrial & domestic (use prohibited 1989)
Endosulfan	Mainly diffuse	Agriculture and forestry (use severely restricted)
DDT	Mainly point	Use as insecticide prohibited in the EU (UK from 1984)
Lindane (Gamma-HCH)	Mainly diffuse	Agriculture, insecticide and public health
Pentachlorophenol	Point and diffuse	Wood preservative
OTHER PESTICIDES		
Atrazine	Mainly diffuse	Agriculture (use prohibited for non-agricultural use 1993)
Simazine	Mainly diffuse	Agriculture (use prohibited for non-agricultural use 1993)
Trifluralin	Mainly diffuse	Agriculture
NUTRIENTS		
Ammonia (NH_3)	Point and diffuse	Domestic and agriculture
Nitrate (NO_3)	Point and diffuse	Domestic and agriculture
Nitrite (NO_2)	Point and diffuse	Domestic and agriculture
Total Oxidised Nitrogen	Point and diffuse	Domestic and agriculture
Total Nitrogen	Point and diffuse	Domestic and agriculture
Orthophosphate	Point and diffuse	Domestic and agriculture
Total Phosphorus	Point and diffuse	Domestic and agriculture
ORGANOMETALS		
Tributyltin	Mainly diffuse	Antifouling paints, wood preservation
Triphenyltin	Mainly diffuse	Used in the synthesis of biocides
Total Organic Tin	Mainly diffuse	Antifouling, wood preservation, biocides

Chapter 3

WHY THIS INFORMATION IS REQUIRED

This section summarises the UK's obligations to monitor and control the discharge of hazardous and nutrient substances to water. These obligations originate from several international agreements and commitments. For example:

- the Paris Convention;
- the North Sea Conference Declarations;
- EC Directives.

3.1 PARIS CONVENTION

The Convention for the Prevention of Marine Pollution from Land-Based Sources (ie. the Paris Convention), was adopted on 4 June 1974 and was brought into effect on 6 May 1978 by the following contracting countries: Belgium, Denmark, France, Germany, Iceland, Ireland, Netherlands, Norway, Portugal, Spain, Sweden and the UK (Figure 3.1). The EU is also a contracting party to the Convention. The area covered by the Convention is the North East Atlantic including the North Sea, but excluding the Baltic and the Mediterranean.

Figure 3.1: Paris & Oslo Commission Countries

Members: Belgium, Denmark, Finland, France, Germany, Iceland, Ireland, Netherlands, Norway, Portugal, Spain, Sweden, United Kingdom.

Table 3.1 PARCOM Black and Grey list substances

BLACK LIST

- Organohalogen compounds and substances which form such compounds in the marine environment, excluding those which are biologically harmless, or which are rapidly converted in the sea into substances which are biologically harmless.

- Mercury and mercury compounds.

- Cadmium and cadmium compounds.

- Persistent synthetic materials which may float, remain in suspension or sink, and which may seriously interfere with any legitimate use of the sea.

- Persistent oils and hydrocarbons of petroleum origin.

GREY LIST

- Organic compounds of phosphorus, silicon and tin and substances which may form such compounds in the marine environment, excluding those which are biologically harmless, or which are rapidly converted in the sea into substances which are biologically harmless.

- Elemental phosphorus.

- Non-persistent oils and hydrocarbons of petroleum origin.

- The following elements and their compounds:

  Arsenic   Chromium   Copper   Lead   Nickel   Zinc

- Substances which have been agreed by the Commission as having deleterious effect on the taste and/or smell of products derived from the marine environment for human consumption.

The Convention introduced two lists of substances for control; the 'Black' list and the 'Grey' list. Contracting countries are obliged to eliminate pollution by substances on the Black list and limit pollution by substances on the Grey list. (See Table 3.1 for details of the two lists.) This approach was later adopted in EC legislation which also defines List I (Black list) and List II (Grey list) substances (see Section 3.3).

The Convention is administered by the Paris Commission (PARCOM) which:

i)   formulates policy to eliminate or reduce existing pollution;

ii) seeks to prevent further contamination of coastal waters or open sea;

iii) requires marine environmental monitoring to be carried out;

iv) assesses the effectiveness of pollution reduction measures;

v)   reports on the results of monitoring.

The Commission is a meeting of representatives from the Contracting Parties. Any measures the Commission wishes to introduce are 'recommendations' which have to be ratified by the contracting countries. The 'recommendations' are not legally binding in the UK, but the UK Government has indicated that it will take steps to implement those recommendations that it accepts. The Paris Convention also provides for the adoption of measures ("Decisions") which are regarded by the United Kingdom as binding in international law if the United Kingdom has voted for them or accepted them.

Table 3.2 Substances for which the UK informs the Paris Commission of discharges to the sea, the 'PARCOM List'

Mercury, Hg	PCBs (the following congeners: IUPAC Nos. 28, 52, 101, 118, 138, 153, 180)	Orthophosphate - PO_4-P
Cadmium, Cd		Total nitrogen - N
Copper, Cu	Gamma HCH (Lindane)	Total phosphorus - P
Zinc, Zn	Nitrate - NO_3-N	Suspended solids
Lead, Pb		

To date, PARCOM has adopted measures to reduce the discharge to sea of several hazardous substances: mercury; cadmium; polychlorinated biphenyls (PCBs); and biocides in general. It has also recommended that inputs of nutrients are reduced in areas where nutrients are the likely cause of eutrophication.

In order to monitor progress in achieving these reductions, PARCOM requires that it be supplied with regular information about inputs of certain substances to see whether those inputs are via rivers or are directly from industrial and sewage discharges. This list of substances, the PARCOM list, is given in Table 3.2.

In 1987, PARCOM recognised that the input data submitted by the Contracting Countries were incomplete and were based on non-standard methodologies. Therefore, at its meeting in Lisbon in June 1988, PARCOM decided to implement a comprehensive annual study of selected pollutants to Convention waters using standard methodologies. The first study using such standard methods was carried out during 1990. The alternative methodologies provided do not, however, provide results which can be directly compared with each other. There were also shortcomings in the application of the methodologies by some of the Contracting Parties. There was no survey undertaken in 1989 as the methods were still being developed. The mandatory parameters to be monitored were those listed in Table 3.2 together with salinity.

The results of the 1990 survey (PARCOM 1992) represented the most complete study of gross riverine and direct inputs to the Convention area up to that time. Nevertheless, the results over the Convention area as a whole were partial and not capable of direct comparisons. The survey objective was to identify at least 90% of the inputs of each

Table 3.3 Industrial sectors that have been reviewed or are under review by PARCOM

Energy production from fossil fuel	Pulp and paper industry
Fertiliser production	Refineries
Foundries	Secondary iron and steel industry
Mining	
Non-ferrous metal industry	Shipyards
Pharmaceutical industry	Surface treatment of metals
Organic chemicals	Tanneries
Primary aluminium industry	Phosphogypsum fertilisers
Primary iron and steel industry	Textile industry
Production and formulation of pesticides/biocides	Waste incineration

Figure 3.2: Annex 1A Countries

Members: Belgium, Denmark, France, Germany, Netherlands, Norway, Sweden, Switzerland, United Kingdom.

Note: France regards its non-North Sea territory as not covered by the Commitment

substance specified in the PARCOM list (Table 3.2). As any one year is unlikely to be average, several years data will be needed to get a fully representative picture. Consequently the survey has been repeated annually since 1990. PARCOM is now assessing:

i) whether annual reporting should continue;

ii) whether the 1990 methodology requires any modification.

In line with moves within the UK and the EU, PARCOM is moving towards an 'industry' rather than 'substance' related approach to controlling discharges. In this case industries which discharge the most polluting substances are targeted and the processes used by those industries reviewed to establish the best available techniques (BAT). These are the technologies which give the lowest practicable output of the substances at a reasonable cost. The industrial sectors currently being studied by PARCOM are listed in Table 3.3.

As well as requiring Contracting Countries to monitor inputs, the Paris Convention also requires them to monitor the marine environment. In September 1992 the member states for the Paris Commission, the Oslo Commission (which deals with direct, non land-based discharges) and the European Union negotiated a new Convention for the Protection of the Marine Environment in the North East Atlantic. When ratified by the member Governments it will replace the Oslo and Paris Conventions which had been ratified by the Governments of Belgium, Denmark, Finland, France, Germany, Iceland, Ireland, Luxembourg, Netherlands, Norway, Portugal, Spain, Sweden, Switzerland and United Kingdom. The new Convention stresses the importance of:

- the precautionary principle;
- the polluter pays principle;
- and the concepts of the best available technology and of the best environmental practice including, where appropriate, clean technology.

The new Convention has endorsed a plan for future work which includes:

i) the establishment of a quality assessment programme of the marine environment in the maritime area;

ii) the reduction of discharges and emissions of substances which are toxic, persistent and liable to bioaccumulate. Concentrations of these substances should be reduced, by the year 2000, to levels which are not harmful to man or nature, with the aim of their elimination;

iii) the prevention and elimination of pollution caused by the dumping of wastes;

iv) agreement to reduce discharges and emissions of nutrients to those areas where these discharges cause eutrophication;

v) the definition of best available techniques, the best environmental practice and clean technology;

vi) the collection of quantitative data about land based discharges and diffuse sources of hazardous substances and nutrients reaching the maritime area.

3.2 NORTH SEA CONFERENCE DECLARATIONS

By 1984 it was generally perceived (by the public and also by some Governments) that slow progress was being made in reducing pollution of the North Sea. As a result a series of Ministerial Conferences were held attended by the Environmental Ministers of all North Sea riparian countries. The first was held in Germany in 1984; the second in London in 1987; the third at the Hague in 1990; and a fourth is planned in Denmark for 1995. At the end of each Conference the Ministers from the participating countries agree objectives by way of Declarations. These Declarations are not legally binding and it is up to each Government to decide how to achieve the stated objectives.

In the Declaration from the Second Conference in 1987, the need to adopt a precautionary approach in relation to the most dangerous substances (defined as those that are persistent, toxic and liable to bioaccumulate) was a major feature. Amongst other issues the Ministers agreed to take measures to:

- reduce the input loads of dangerous substances, from rivers and estuaries, to the North Sea by around 50% by 1995 using input load in 1985 as a baseline;
- reduce inputs of phosphorus and nitrogen by around 50% between 1985 and 1995 into areas where these inputs are likely to cause pollution.

For the dangerous substances, each country had to draw up its own priority list from a reference list of 170 substances. In the UK a list of 23 substances was created based on their toxicity, persistence, bioaccumulation and volume of usage: this list is commonly

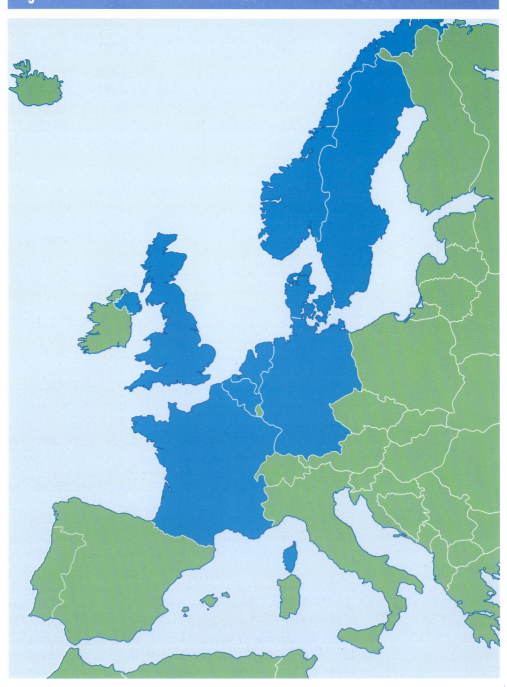

Members: Belgium, Denmark, France, Germany, Netherlands, Norway, Sweden, United Kingdom.

referred to as the 'Red List' (see Table 3.4). Inputs from point sources (eg. industrial discharges) were to be reduced through the application of best available technology (BAT). Diffuse source inputs (eg. inputs from run-off of pesticides from agricultural land) were to be reduced by controls on supply, use and disposal of products.

At the Third Conference in 1990, Ministers agreed a common list of 36 dangerous substances, referred to as the North Sea Conference Common, or Annex 1A, List (see Table 3.4 and Figure 3.2). They also reiterated their previous commitments concerning inputs of hazardous substances and agreed to:

- achieve significant reductions of these 36 substances, from rivers and estuaries, to the North Sea by around 50% between 1985 and 1995;
- to reduce the total inputs, from all sources, of dioxins, mercury, cadmium and lead by around 70% or more between 1985 and 1995 (provided that the use of best available technology or other low waste technologies enable such reductions);

- to reduce the input of nutrients, by around 50%, between 1985 and 1995 in areas where these inputs are likely to cause pollution;

- to make substantial reductions in the quantities of pesticides reaching the North Sea with special attention to phasing out those which are the most persistent, toxic and liable to bioaccumulate;

- to phase out and destroy all identifiable PCBs by 1999.

Further reductions in inputs are required by the year 2000 to a point where inputs of hazardous substances no longer represent a danger to man or nature.

The area of sea affected by the North Sea Declarations is confined to the North Sea and Channel and immediately adjoining waters connecting it to the Baltic. The Government has set out its intentions with regard to these commitments in a Guidance Note issued after the 1990 North Sea Conference. One point from the Note is that the Government is applying the policies to all UK coastal waters and not just the North Sea.

The UK Government announced in 1990 that a programme to monitor riverine inputs to the sea was to be introduced in line with the recommendations of the Paris Commission. The objective of the programme was to monitor on an annual basis at least 90% of all inputs of each substance via rivers at tidal limits, and via sewage, sewage effluent or direct industrial discharges downstream of tidal limits (DoE 1990).

Following the London Declaration in 1987, the North Sea Task Force (NSTF) was set up (Figure 3.3) during 1988 with the following membership; Belgium, Denmark, France, Germany, Netherlands, Norway, Sweden and the UK. The primary objective of the NSTF was to produce a new North Sea Quality Status Report (QSR) by the end of 1993, to update the initial 1987 QSR. The NSTF created an extensive monitoring network, providing the various countries bordering the North Sea with a responsibility for specific areas. The intention of this division into areas was to establish the effects of the measures introduced by the London Declaration and also to identify additional measures that might be required.

The NSTF recognised that in order to put together a more comprehensive document than had been produced in the past, it would have to gather more comparable data on substance concentrations and their effects. Additionally, it realised that to produce such a report by 1993, any new initiatives would have to be put into place quickly. The NSTF agreed a Monitoring Master Plan (MMP) in 1989 for implementation during the 1990/91 PARCOM surveys, which addressed the issues raised by the London Declaration in 1987. Each North Sea member carried out a field monitoring programme at sea and in coastal waters and the data have now been published during 1994 in the QSR.

3.3 EC DIRECTIVES

On the 4 May 1976, the Ministers of the EC adopted a Directive (76/464/EEC) on "pollution caused by certain dangerous substances discharged into the aquatic environment". It is commonly referred to as the "Dangerous Substances Directive". The Directive followed on from the approach adopted in the Paris Convention and established two lists of compounds, List I (also known as the Black list) and List II (also known as the Grey list).

List I substances are regarded as particularly dangerous because of their toxicity, persistence and bioaccumulation. Pollution by List I substances must be eliminated. The EC lays down standards for these substances in 'Daughter Directives'. List II substances are less dangerous but may still have a deleterious effect on the aquatic environment. Pollution by List II substances must be reduced. The EC Member States set standards for these in national law.

In 1982, the European Commission published a list of 129 potential List I substances (subsequently extended to 132). These substances are being assessed by the European

Commission and so far "Daughter Directives" have been adopted by the Council of Ministers identifying 17 List I substances (Table 3.4).

All discharges of List I substances must comply with standards set in the 'Daughter Directives'. There are two types of standards, limit values and environmental quality objectives (equivalent to environmental quality standards in the UK), and Member States can choose which of these to use. For List II substances Member States should use the environmental quality objective approach to control discharges.

The monitoring required for EC Directive compliance is discussed no further as this report seeks to concentrate on the survey work carried out for the Paris Commission and North Sea Conference purposes. The recently adopted UK Integrated Pollution Control (IPC) procedure is dealt with in Chapter 8.

Table 3.4 U.K. and North Sea Monitoring Requirements

	List I	List II	Red List	Annex 1A	Parcom	NSTF
METALS						
Mercury (Hg)	****		****[1]	****	****	****
Cadmium (Cd)	****		****[1]	****	****	****
Copper (Cu)		****		****	****	****
Zinc (Zn)		****		****	****	****
Lead (Pb)		****		****	****	****
Tributyltin (TBT)			****	****		
Triphenyltin (TPT)			****	****		
Organotins		****				
Chromium (Cr)		****		****		****
Nickel (Ni)		****		****		****
Arsenic (As)		****		****		****
Boron (B)		****	****	****	****	****
Vanadium (V)		****				
PCBs						
PCBs			****			****
PCB 28					****	
PCB 52					****	
PCB 101					****	
PCB 118					****	
PCB 138					****	
PCB 153					****	
PCB 180					****	
PESTICIDES & ORGANOCHLORINES						
Hexachlorocyclohexane	****			****		****
Gamma-HCH (Lindane)			****		****	
DDT	****		****	****		
Drins				****		****
Aldrin	****		****			
Dieldrin	****		****			
Endrin	****		****			
Isodrin	****					
Trifluralin			****	****		
Trichlorobenzene	****		****	****		
Trichloroethylene	****			****		

Note: [1] all Compounds

Table 3.4 U.K. and
North Sea Monitoring
Requirements,
continued

	List I	List II	Red List	Annex 1A	Parcom	NSTF
Tetrachloroethylene	✳✳✳✳			✳✳✳✳		
Hexachlorobenzene	✳✳✳✳		✳✳✳✳	✳✳✳✳		
Hexachlorobutadiene	✳✳✳✳		✳✳✳✳	✳✳✳✳		
Carbon Tetrachloride	✳✳✳✳			✳✳✳✳		
Chloroform	✳✳✳✳			✳✳✳✳		
Endosulfan			✳✳✳✳	✳✳✳✳		
Dichlorvos			✳✳✳✳	✳✳✳✳		
Fenitrothion			✳✳✳✳	✳✳✳✳		
Fenthion				✳✳✳✳		
Malathion			✳✳✳✳	✳✳✳✳		
Parathion				✳✳✳✳		
Parathion-methyl				✳✳✳✳		
Azinphos-ethyl				✳✳✳✳		
Azinphos-methyl			✳✳✳✳	✳✳✳✳		
Atrazine			✳✳✳✳	✳✳✳✳		
Simazine	✳✳✳✳		✳✳✳✳	✳✳✳✳		
Pentachlorophenol	✳✳✳✳		✳✳✳✳	✳✳✳✳		
1,2 Dichloroethane			✳✳✳✳	✳✳✳✳		
Trichloroethane				✳✳✳✳		
Mothproofing Agents		✳✳✳✳				
Dioxins				✳✳✳✳		
NUTRIENTS						
Ammonia						✳✳✳✳
Nitrate					✳✳✳✳	
Nitrite						✳✳✳✳
Total Oxidised Nitrogen						✳✳✳✳
Total Nitrogen					✳✳✳✳	
Orthophosphate						
Total Inorganic Phosphate					✳✳✳✳	
Total Phosphorus					✳✳✳✳	✳✳✳✳
Silica						✳✳✳✳
MISCELLANEOUS						
Suspended Particulate Matter					✳✳✳✳	✳✳✳✳
Temperature						✳✳✳✳
Salinity					✳✳✳✳	✳✳✳✳
pH		✳✳✳✳				
Dissolved Oxygen						✳✳✳✳

Chapter 4   HOW LOADS ARE CALCULATED

Loads are the product of the contaminant concentration and the river or effluent flow. PARCOM has provided standard methods for the estimation and calculation of input loads to coastal waters, and the NRA has adopted these recommendations in reporting the results of its sampling programmes for PARCOM and Annex 1A purposes.

All the main English and Welsh river systems are generally sampled monthly at a sampling point close to, but upstream of, the tidal limit for a wide range of contaminants. In addition, all major direct discharges of trade or sewage effluent entering downstream of that sampling point are sampled, as are major coastal discharges. The river sampling regime is designed to cover the whole flow cycle but, where possible, with a bias towards periods of expected high river flow, when the loads carried are expected to be higher. Obtaining this bias was not possible in 1990 and 1991 as these years were drier than most. The aim is to sample all those rivers and effluents that contribute significantly to the total input load of identified substances. It is impossible to sample 100% of the total input load as the final few percent of that load are inevitably spread in tiny amounts across large numbers of very small streams or effluents. The PARCOM recommendations allow for this and suggest the aim should be to sample 90% of the total load. For rivers that are relatively unpolluted, sampling is reduced in line with PARCOM's recommendations to:

- **4 times a year for rivers where the input of substances does not contribute significantly to 90% of the total input load;**
- **once a year for rivers where the specified substances are at or below the detection limit of the analytical methods used.**

As a result of this rationalisation the NRA analysis of a river sample can vary month by month. Those substances which are found at high concentrations in the river will be analysed for each month. Those substances which are found at lower concentrations in the river will only be analysed for every third month or every twelfth month.

The analytical effort "saved" by this rationalisation is used to examine new discharges or previously unsampled discharges to ensure that no significant sources of contaminants are being missed. As the years go by this rationalisation and redeployment of analytical effort will gradually improve the reliability of the monitoring programme. This system is considered to be effective in enabling the NRA to identify at least 90% of the total input to the coastal areas of England and Wales.

PARCOM gives two methods for calculating river loads: one based on mean annual flow rates (for which river flow on each day of the year is needed) and one using average actual loads on the sampling days (for which river flow is only needed on each sampling day). The NRA uses the second method as data on annual mean river flows are not readily available. However, a comparison of the two methods on the River Thames, for which mean annual flow rates were available, indicated that the methods gave comparable results.

As part of the PARCOM standard methods, limits of detection (LOD) for the chemical analysis of certain contaminants were also given. The numeric value of each limit of detection was chosen to try to ensure that a substantial majority (ie. at least 70%) of all samples analysed would leave results above the limit of detection. Whilst it is inevitable that some samples generate results below the limit of detection, such results are difficult to interpret (see Box 4.1). It is sensible to try to minimise the numbers of results which are below the limit of detection. The following were recommended as suitable detection limits:

- **mercury and cadmium**   10 ng/l
- **copper and lead**   100 ng/l
- **PCBs (for each selected congener)**   1 ng/l
- **zinc**   500 ng/l
- **gamma-HCH (lindane)**   1 ng/l

Box 4.1: Limits of Detection

When a substance is only present in a sample in very small amounts, it may not be detected by even the most sophisticated chemical analysis. The break point at which any analytical method can just 'see' that the substance is present in the sample is called the limit of detection. Any amount smaller than this limit is indistinguishable from a zero amount and cannot be quantified.

Rather than record all such amounts as zero (which would not be true) it is usual to refer to them as being less than the numeric value of the limit of detection; eg. <10ng/l etc. Thus in this example any true value between 0ng/l and 10ng/l would not be seen by the analytical method and all such values would be recorded as <10ng/l.

The only positive statement that can be made about such results is that if the substance is present at all in the sample, it cannot be present at a concentration greater then 10ng/l.

Limits of detection can vary between laboratories due to differing instrumentation and analytical methodology. They can also vary within the same laboratory over time for the same reasons. This can effect load calculations where some of the concentration results used are less than values. Thus, some of the fluctuations in load figures over the years will be due to the analytical variability rather than actual increases or decreases in those loads.

In all cases the total concentration is determined, rather than the dissolved portion or that associated with particulate matter in the water. The NRA has adopted these LODs, where possible, and in some cases for river samples has improved upon them (eg. 5ng/l mercury).

Two sets of load estimates are requested and supplied to PARCOM. The first treats results recorded as less than the limit of detection as having a true concentration of zero. The second set treats such results as having a true concentration at the limit of detection. The first is considered to be the lower estimate of load and the latter as the upper estimate. If most results are below the limit of detection there can be large differences between the two. Hence, it is useful to have limits of detection as low as is practical to achieve.

The North Sea Conference Declarations do not specify how loads should be calculated and so the NRA has chosen to use the same procedure it uses for Paris Commission purposes.

Loads, being a product of two numbers, are subject to an inherent variability. This can be particularly noticeable when calculating river loads. River flows vary significantly. A winter flood flow can be 100 times greater than a dry weather summer flow. It is difficult to have a flow gauging mechanism that can cope with these extreme flows and yet measure both accurately. Also contaminant concentrations in river water can be very low, close to or below the limit of detection; the point where there is least confidence with the analytical result. The multiplication of a very large imprecise number by a very small imprecise number must give a product which itself is of limited validity. Thus, for rivers where large flows and low contaminant concentrations predominate, load calculations inevitably result in answers that are only a broad estimate of true load.

Load calculations for discharges of effluent (industrial or sewage) are also subject to an inherent variability. Flow variation is usually less extreme than for a river, and contaminant concentrations are usually higher. Thus, both components of the load calculation can often be measured for an effluent with more precision than for a river water. Hence, a better estimate of true load is often obtained for an effluent than is possible for a river water.

Box 4.2: High and Low Loads

Where a number of concentration results are being added together prior to calculating an average concentration, and some of those concentrations are recorded as less than values (see Box 4.1), there is no mathematically correct way of adding in those less than values, nor any way of calculating a mathematically correct average concentration value. The best that can be done is to calculate two averages. For the first, all less than values are ignored; ie. they are treated as being zero. For the second, all less than values are assumed to have the full numeric concentration of the less than value; ie. 10ng/l for a <10ng/l value. The true average will lie somewhere between the two averages: but it cannot be below the lower average; nor above the higher.

Where loads are being calculated (ie. average concentration is being multiplied by average flow to give a product in, say, kg/year) and concentration is given as a higher and lower figure due to some of the contributing concentrations being less than values, then no true single load figure can be calculated. One way round this is to calculate two load figures: one using the higher average concentration to give a high load figure; and one using the lower average concentration to give a low load figure. The true load will, therefore, lie somewhere between the low load and the high load. All that can be said with certainty is that the actual load cannot be less than the low load but that it could be as high as the high load (although the latter is unlikely). Thus, both loads must be looked at before we can know the degree of confidence we can have in any decision.

EXAMPLE

A metal contaminant in river water has a limit of detection of 1ug/l. The river has an annual average flow of 100 m³/sec. If all the samples analysed for the metal are below the limit of detection (ie. all are <1ug/l) then the low and high loads for the metal are 0kg/year and 3154 kg/year. Thus the true load the river is carrying is somewhere between 0kg and 3154kg each year.

Any decision which involves adding an extra load to the river of, say, 500kg per year (or removing 500kg per year from the existing load) would be of very uncertain validity. It would only be decisions involving loads of about 3000kg per year or more where it would be possible to be reasonably certain that a measurable effect would result.

As rivers generally have large flows and low contaminant concentrations a single atypical concentration can have a dramatic effect on annual load figures. In the example given above, if out of ten samples all are below the limit of detection except one which is, say, 10ug/l, then the low load rises to 3154 kg/yr and the high load to 6000 kg/yr. The larger the river flow the more dramatic the effect an atypical result will have. Here even a decision involving a load of 3000 kg/yr would be of limited validity.

These points need to be borne in mind when considering load data particularly for the country's larger rivers.

Where load discharged to a river catchment or to a coastal sea zone is to be quoted the question arises whether the high or low load should be used. There is no right answer as neither load is wholly correct. However, any concern which may be felt at the size of a low load has to be reinforced by the knowledge that the true load will be higher still. Thus, there is some advantage in looking at low loads first and this is the NRA's usual course of action. Ultimately it is only by looking at both loads and at how much of the concentration data is above the limit of detection that understanding can be as complete as possible.

Chapter 5

TRENDS SINCE 1985

One of the main objectives of the North Sea Declarations is the commitment to reduce inputs of dangerous substances to the North Sea. Within England and Wales significant progress has been made in reducing inputs. Current loads for those substances for which 1985 baseline data are available (principally the Black and Grey list metals) are mostly at or below 50% of their 1985 values. Further reduction measures are in hand for some of these substances which will come to fruition over the next year or two.

Table 5.1 gives details of the 1985 baseline loads of most of the metals on the Annex 1A list and of lindane (as provided by the DoE from its UK North Sea Action Plan). Also given are the loads as measured by the NRA for the years 1990 to 1993. Figures 5.1 and 5.2 present this information graphically together with data for chromium and nickel and show what progress has been made against the 1995 target for a reduction of about 50%. The target for cadmium, mercury and lead is now for a 70% reduction, but this is for all input routes including emissions to air.

High loads (see Box 4.2 in previous chapter) have been used in presenting this data, as only high load data is available for 1985. The data comes from the NRA's Paris Commission monitoring except for chromium and nickel which are not included on the PARCOM list (see Table 3.2). The data for these two metals comes from the NRA's Annex 1A monitoring programme.

The North Sea Conference Declarations require inputs to the English Channel to be included in North Sea data returns. Thus in Table 5.1 and Figures 5.1 and 5.2 data from both the North Sea and Channel coastal seas have been combined. Data from all four English and Welsh coastal seas (see Box 6.1) have also been combined and are presented for comparison purposes.

For many of the Annex 1A substances no comprehensive 1985 data exists and for these substances no reliable 1985 baseline can be established. Thus load reductions against a 1985 baseline cannot be quantified, although estimates can be made.

Table 5.1 High loads (tonnes/yr) arising from riverine and direct inputs to the North Sea/Channel and all English and Welsh coastal seas between 1985 and 1993

		1985	1990	1991	1992	1993	1985-1993 % Reduction
Cd	North Sea/Channel	23.6	12.5	14.4	7.8	6.7	72
	Eng & Wal	64	46.2	39.3	29.1	22.2	65
Hg	North Sea/Channel	7.3	3.6	3.0	2.1	2.0	73
	Eng & Wal	24.6	9.1	6.4	5.4	4.9	80
Cu	North Sea/Channel	398	325	293	312	275	31
	Eng & Wal	1098	499	506	482	439	60
Zn	North Sea/Channel	1760	1810	1780	2320[1]	1730	2
	Eng & Wal	3340	3130	3100	3300[1]	2630	21
Pb	North Sea/Channel	420	205	337	257	342[1]	19
	Eng & Wal	730	427	465	405	532[1]	27
Lindane	North Sea/Channel	0.39	0.30	0.32	0.22	0.26	33
	Eng & Wal	1.38	0.43	0.65	0.44	0.51	63

[1] These estimates include large atypical inputs

Since 1991 the NRA has been collecting data annually for all Annex 1A substances. Data from 1991 and 1992 could be used (for those Annex 1A substances for which no previous data exists) to create a baseline against which loads in future years could be compared to establish trends. However, this may not be straightforward. The 1991-1993 data for these substances shows extreme variability with many analytical results being at, or below, the limit of detection. The high and low load figures (see Box 4.2) can differ markedly and it maybe difficult to discern any reliable trend from the data for some years to come. Meanwhile, if evidence of load reductions is needed, it will have to come from other sources: for example from records of amounts used, or of increasing restrictions on the applications for which pesticides are approved.

There are other national and international initiatives which will have some effect on contaminant loads discharging into coastal seas. For example the Urban Waste Water Treatment Directive will result in some estuarine and coastal sewage discharges being treated more extensively than at present. Similarly the gradual implementation of Integrated Pollution Control (IPC) will result in some industrial discharges receiving more treatment.

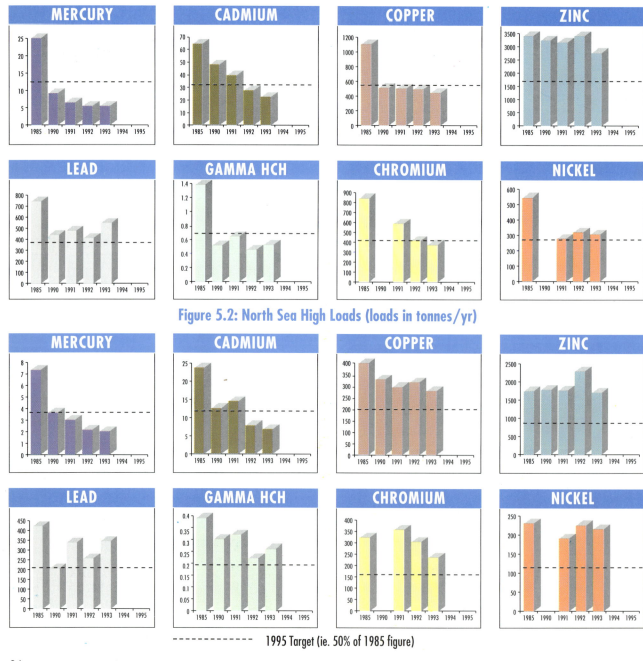

Figure 5.1: England & Wales High Loads (loads in tonnes/yr)

Figure 5.2: North Sea High Loads (loads in tonnes/yr)

- - - - - - - - - - - - - - - - 1995 Target (ie. 50% of 1985 figure)

Chapter 6

RESULTS OF THE 1990-93 SURVEYS

This section describes the results of the NRA's 1990 - 1993 Paris Commission and North Sea Conference Annex 1A monitoring. (Tables summarising the results for all contaminants are available as an Appendix from TaPS Centre, Anglian Region). Results for the main contaminants (the PARCOM list metals; lindane; simazine; and the nutrients, total nitrogen and phosphorus) have been abstracted and are assessed within this section in three ways:

- as loads to coastal regions by input source (ie. rivers, industrial effluents and sewage effluents) in 1990 - 1993, for England and Wales;
- as loads from eight major estuaries for 1990 - 1993, in England and Wales;
- by location of the 'top ten' loads in 1990 - 1993, from the area covered by the NRA ie. England and Wales;

For these comparisons a 'low load' figure is used (see Box 4.2 in Chapter 4).

Box 6.1: Sea Regions Definition

Data have often been summarised by relatively small zones known as ICES (International Council for the Exploration of the Sea) regions. These ICES regions have been grouped together to form four regions: North Sea; English Channel; Bristol Channel; & Irish Sea.

For North Sea Declaration purposes the English Channel (as far west as 5° W) is regarded as the North Sea. To avoid confusion no attempt has been made in the text of this chapter to use this larger definition of the North Sea. Thus, the four coastal seas are as shown in Figure 6.1.

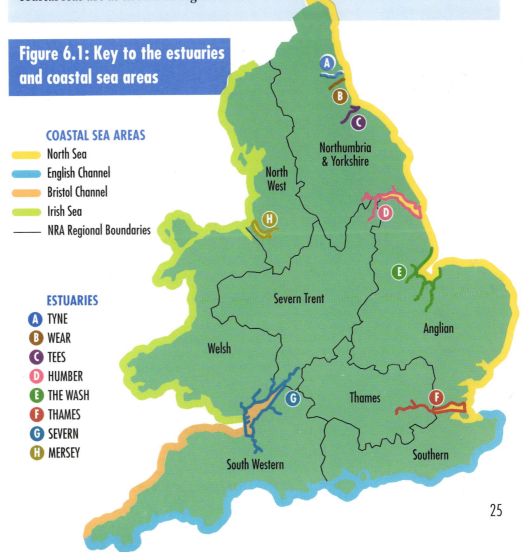

Figure 6.1: Key to the estuaries and coastal sea areas

COASTAL SEA AREAS
- North Sea
- English Channel
- Bristol Channel
- Irish Sea
- NRA Regional Boundaries

ESTUARIES
- A TYNE
- B WEAR
- C TEES
- D HUMBER
- E THE WASH
- F THAMES
- G SEVERN
- H MERSEY

North West

Northumbria & Yorkshire

Severn Trent

Welsh

Anglian

Thames

Southern

South Western

CADMIUM

In 1990 industrial inputs, particularly to the Irish Sea, were by far the largest contributors of cadmium to coastal waters. However, there have been substantial reductions in industrial inputs since 1990. By 1993, although the Irish Sea was still receiving the largest load, overall the total load was halved.

Of the major estuaries in England and Wales the Severn receives the largest input of cadmium, although this is now decreasing. This originates from the metal smelting industry on Severnside. The fluctuation between riverine input and industrial input as the major load is not real. It merely reflects the uncertain status of one large input. Legal opinion is being sought to resolve the matter. The Humber is the other estuary receiving significant cadmium inputs. Here there has been a marked decline in industrial input from smelting and fertiliser manufacture, but river based loads have risen. In part this rise in river loads mirrors the increased river flows following the drought years of 1990 and 1991, but further investigation of other reasons is in hand.

Figure 6.2: Annual Cadmium Low Load input into Coastal Waters (England & Wales)

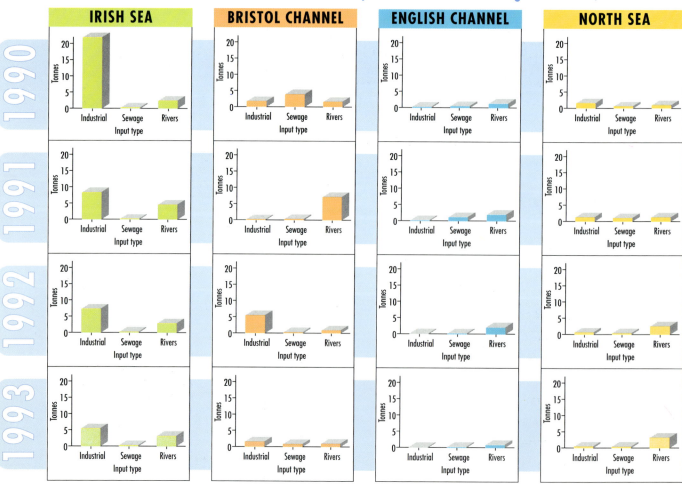

(For explanation of the fluctuation in the Bristol Channel see above)

Figure 6.3: Annual Cadmium Discharges into Major Estuaries 1990-1993

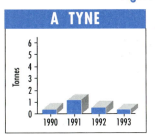
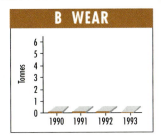
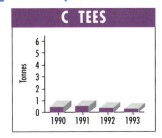
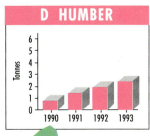

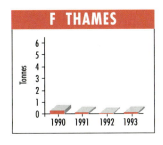
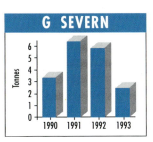
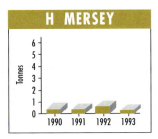

Figure 6.4: Leading Cadmium Discharges

The biggest single source of cadmium is an industrial input on the Cumbrian coast. The next nine largest single sources for each year are shown on Figure 6.4 and detailed in Table 6.1.

Table 6.1: Cadmium Discharges 1990-93, ranked by size as a percentage of the total load measured in each year

No.	Input Type	NRA Region	Catchment	1990 Rank	1990 %	1991 Rank	1991 %	1992 Rank	1992 %	1993 Rank	1993 %
	TOTAL (Tonnes)			**36.7**		**26.6**		**21.4**		**15.8**	
1	Industrial	North West	Irish Sea	1	60.6	1	30.8	1	33.3	1	32.5
2	Sewage	South Western	Severn Estuary	2	9.2	2	23.2	2	15.1	7	3.2
3	Industrial	South Western	Severn Estuary	3	4.1			3	9.2	3	8.7
4	River	South Western	Fal (Restronguet Cr)	4	2.0			4	5.4		
5	River	North West	Ribble	5	1.8	5	2.5	9	1.9	2	9.0
6	River	Welsh	Usk	6	1.6						
7	Industrial	Northumb/Yorks	Humber	7	1.4	9	1.5				
8	River	Welsh	Wye	8	1.3						
9	Industrial	Northumb/Yorks	North Sea	9	1.1						
10	River	North West	Mersey	10	0.8			10	1.4		
11	River	Welsh	Conwy			3	3.9				
12	River	North West	Eden			4	3.5	6	3.0	10	1.6
13	Sewage	Northumb/Yorks	Tyne			6	2.3				
14	River	Northumb/Yorks	Tyne			7	1.7	8	2.2	9	2.0
15	River	Severn-Trent	Trent			8	1.6	5	4.7	4	5.2
16	River	Severn-Trent	Severn					7	2.2	6	3.2
17	River	Northumb/Yorks	Ouse							5	5.0
18	River	Northumb/Yorks	Ouse (Don)							8	2.1
19	River	North West	Derwent			10	1.4				

27

MERCURY

The biggest single input of mercury is from an industrialised source on Merseyside. Most of the mercury inputs arise in northern England. Whilst in part this is due to the concentration of heavy industry in north west England, some is also likely to be of geological origin.

The Irish Sea is the sea area receiving the greatest mercury loads, with industrial sources being predominant in 1990-1992. By 1993 inputs from rivers were predominant, due partly to decreasing industrial loads and partly increasing river loads as flows increased with the wetter weather.

The Mersey is the estuary which receives by far the largest input of mercury, although, as can be seen, the total load has dropped substantially since 1990. Of the remaining estuaries only the Tyne and Tees are of any significance.

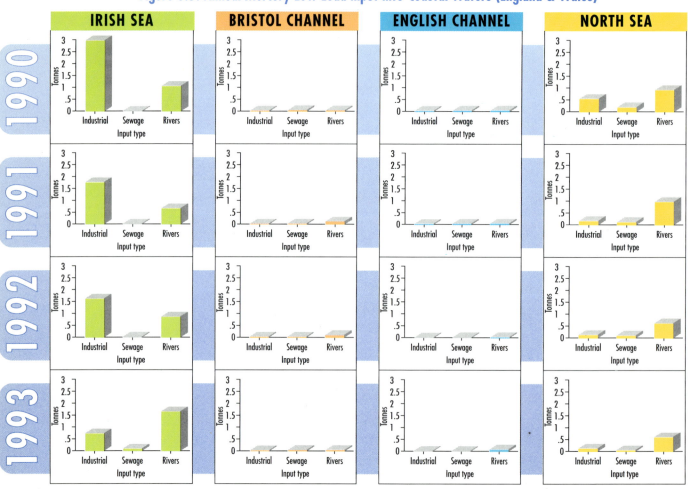

Figure 6.5: Annual Mercury Low Load input into Coastal Waters (England & Wales)

Figure 6.6: Annual Mercury Discharges into Major Estuaries 1990-1993

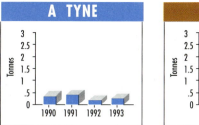

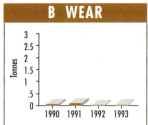

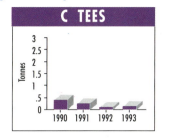

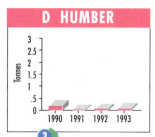

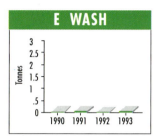

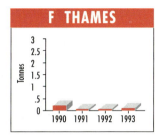

Figure 6.7: Leading Mercury Discharges

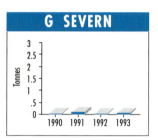

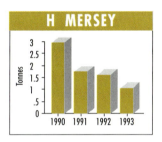

Table 6.2: Mercury Discharges 1990-93, ranked by size as a percentage of the total load measured in each year

No.	Input Type	NRA Region	Catchment	1990 Rank	1990 %	1991 Rank	1991 %	1992 Rank	1992 %	1993 Rank	1993 %
	TOTAL (Tonnes)				**5.8**		**3.9**		**3.4**		**3.1**
1	Industrial	North West	Weaver (Mersey)	1	36.9	1	32.1	1	22.5	1	10.6
2	Industrial	North West	Mersey Estuary	2	8.2	6	5.6	3	10.9		
3	River	Northumb/Yorks	Tweed	3	6.4	3	7.6	5	8.0	10	3.9
4	River	North West	Weaver (Mersey)	4	6.0	4	6.1	4	10.7	5	7.2
5	River	North West	Eden	5	5.4	7	4.6	6	4.8	3	9.6
6	River	Northumb/Yorks	Tyne	6	4.8	2	7.8	8	3.8	6	5.9
7	Industrial	Northumb/Yorks	Tees	7	4.7						
8	River	North West	Lune	8	3.0			10	1.7		
9	Industrial	North West	Wyre Estuary	9	3.0	5	5.7	2	11.3	4	7.7
10	Industrial	Welsh	Milford Haven	10	2.5						
11	River	Northumb/Yorks	Wear			8	2.4				
12	River	Northumb/Yorks	Tees			9	2.4				
13	River	North West	Mersey			10	1.8	7	3.9	8	5.3
14	Industrial	North West	Irish Sea					9	2.6		
15	River	Welsh	Tywi							2	10.3
16	River	North West	Ribble							7	5.7
17	River	North West	Derwent							9	5.0

COPPER

During the period 1990-1993 the total load of copper has remained broadly constant. In 1990 and 1991 the biggest single source was an industrial input to the Tees, but as this reduced in 1992 and 1993 other sources became relatively more important (see Figure 6.9 and Table 6.3).

The North Sea is the sea area which receives the highest loads of copper; over twice the load of that entering any other sea area. Industrial inputs of copper declined significantly during the 1990-1993 period, but riverine inputs have apparently increased although this just maybe a reflection of the wetter weather of 1992 and 1993. Inputs arising from sewage effluent remained steady.

The Humber is the estuary receiving the biggest load of copper; this load has remained almost constant during the 1990-1993 period. The two principal tributary rivers of the Humber, the Trent and the Ouse, are the main sources of the copper. With the exception of the Tees (where loads are now declining from a peak in 1991) the other estuaries do not carry significant loads of copper.

Figure 6.8: Annual Copper Low Load input into Coastal Waters (England & Wales)

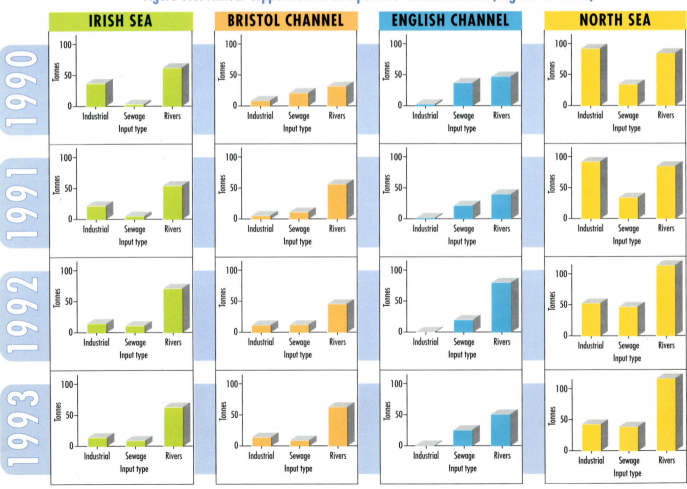

Figure 6.9: Annual Copper Discharges into Major Estuaries 1990-1993

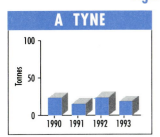

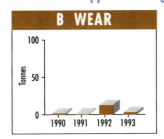

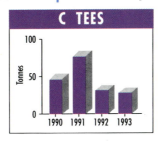

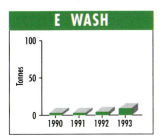

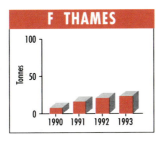

Figure 6.10: Leading Copper Discharges

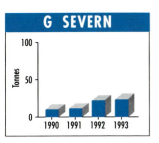

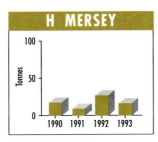

Table 6.3: Copper Discharges 1990-93, ranked by size as a percentage of the total load measured in each year

No.	Input Type	NRA Region	Catchment	1990 Rank	1990 %	1991 Rank	1991 %	1992 Rank	1992 %	1993 Rank	1993 %
	TOTAL (Tonnes)			**468**		**426**		**468**		**418**	
1	Industrial	Northumb/Yorks	Tees	1	9.7	1	17.1	3	5.4	2	4.9
2	River	Northumb/Yorks	Ouse	2	6.0					10	2.6
3	River	South Western	Fal (Restronguet Cr)	3	5.8	6	3.4	1	7.7	3	4.6
4	Industrial	North West	Irish Sea	4	4.3	3	4.3	10	2.4		
5	River	Severn Trent	Trent	5	3.9	2	6.4	2	7.7	1	8.1
6	Industrial	Northumb/Yorks	North Sea	6	3.3						
7	River	Northumb/Yorks	Ouse (Aire)	7	2.6	5	3.5				
8	River	North West	Mersey	8	2.4			9	2.4		
9	Industrial	Anglian	Humber	9	1.9	10	2.1				
10	River	North West	Ribble	10	1.9						
11	River	Welsh	Wye			4	3.6				
12	Sewage	Northumb/Yorks	Tyne			7	2.5	4	4.4	6	3.2
13	River	South Western	Tamar			9	2.1	6	3.2	8	2.7
14	River	Welsh	Afon Goch			8	2.2			9	2.6
15	River	Severn Trent	Severn					5	3.4	4	4.5
16	River	Northumb/Yorks	Tweed					7	2.7		
17	River	Northumb/Yorks	Wear					8	2.6		
18	River	Welsh	Usk							5	3.5
19	River	Northumb/Yorks	Ouse (Don)							7	3.0

31

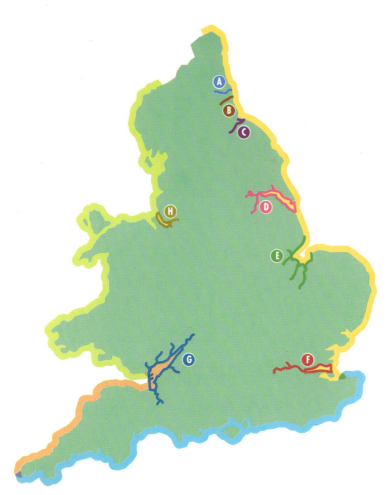

LEAD

Almost all the lead inputs of any size enter via rivers, particularly in the northern part of the country. Much of this is probably due to mining activity, some of it in previous centuries (although there are numerous possible minor point and diffuse sources of lead within river catchments). Lead in petrol is a significant diffuse source of the metal.

The North Sea receives the greatest input of lead, with its tributary rivers being the main source. The increased river flows following the 1990-1991 drought years are thought to be the main reason for the increased loads.

The Humber estuary receives by far the largest load of lead of any of the major estuaries. Most of this enters via rivers and much of this is thought to originate from past mining activity. Due to the historic and diffuse nature of lead contamination within the catchment future remediation is likely to be difficult.

Figure 6.11: Annual Lead Low Load input into Coastal Waters (England & Wales)

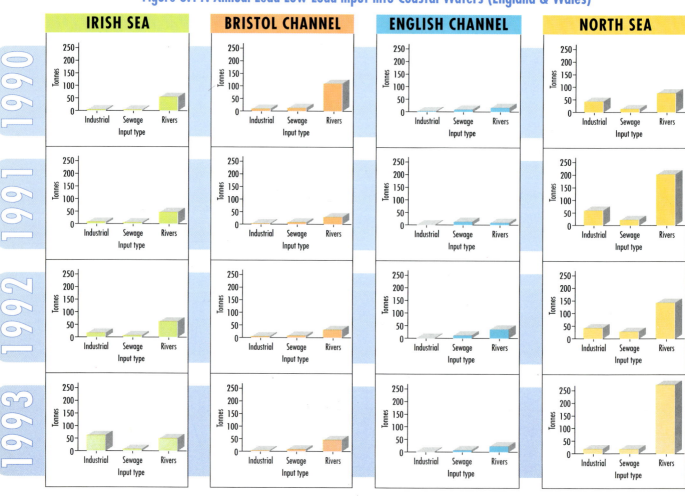

Figure 6.12: Annual Lead Discharges into Major Estuaries 1990-1993

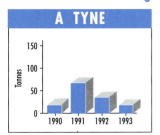

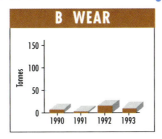

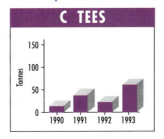

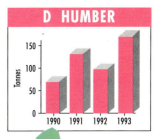

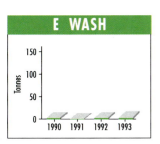

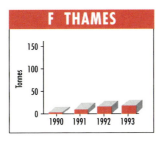

Figure 6.13: Leading Lead Discharges

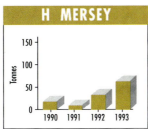

Table 6.4: Lead Discharges 1990-93, ranked by size as a percentage of the total load measured in each year

No.	Input Type	NRA Region	Catchment	1990 336 Rank	1990 336 %	1991 382 Rank	1991 382 %	1992 364 Rank	1992 364 %	1993 488 Rank	1993 488 %
	TOTAL (Tonnes)			336		382		364		488	
1	River	Welsh	Wye	1	13.6						
2	River	Welsh	Usk	2	10.1					8	2.4
3	River	Northumb/Yorks	Ouse	3	8.6	1	18.0	2	7.4	1	22.2
4	Industrial	Anglian	Humber	4	6.4	5	5.4	4	5.1		
5	River	Severn Trent	Severn	5	5.8			5	4.8	6	3.6
6	River	North West	Mersey	6	4.6			10	3.2		
7	River	North West	Ribble	7	4.3						
8	River	Northumb/Yorks	Wear	8	2.3			6	4.1	10	2.1
9	River	Northumb/Yorks	Ouse (Don)	9	2.2					5	4.8
10	River	Severn Trent	Trent	10	2.2	3	5.9	1	9.5	4	5.7
11	River	Northumb/Yorks	Ouse (Wharfe)			10	1.9				
12	River	Northumb/Yorks	Ouse (Aire)			9	2.7				
13	River	North West	Eden			8	2.8	7	3.8	9	2.1
14	River	Northumb/Yorks	Tyne			2	14.7	3	6.5	7	2.9
15	River	Northumb/Yorks	Tees			4	5.6			2	11.4
16	Industrial	Northumb/Yorks	Tees			6	3.5	8	3.4		
17	Industrial	Northumb/Yorks	North Sea			7	2.9				
18	Industrial	North West	Mersey					9	3.3	3	10.5

33

ZINC

Zinc inputs come from a wide range of diffuse sources. The biggest single source of zinc is an industrial input on the south Humber bank. It is noticable that industrial inputs are much more significant for zinc than for the other Grey list metals, copper and lead.

By the end of 1995 the benefit of a waste minimisation programme at the south Humber bank industrial plant should reduce inputs of zinc by about 200 tonnes per annum.

The North Sea receives the highest loads of zinc from England and Wales (reflecting the impact of the Humber), followed by the Irish Sea. The relatively high value for the English Channel in 1992 was due to inputs from abandoned mine workings which have now been reduced.

Of the major estuaries the Humber receives by far the largest load of zinc. This is attributable in the main to one industrial input, although the tributary rivers, the Trent and the Ouse, do carry a significant load. None of the seven remaining major estuaries are important in terms of zinc loadings.

Figure 6.14: Annual Zinc Low Load input into Coastal Waters (England & Wales)

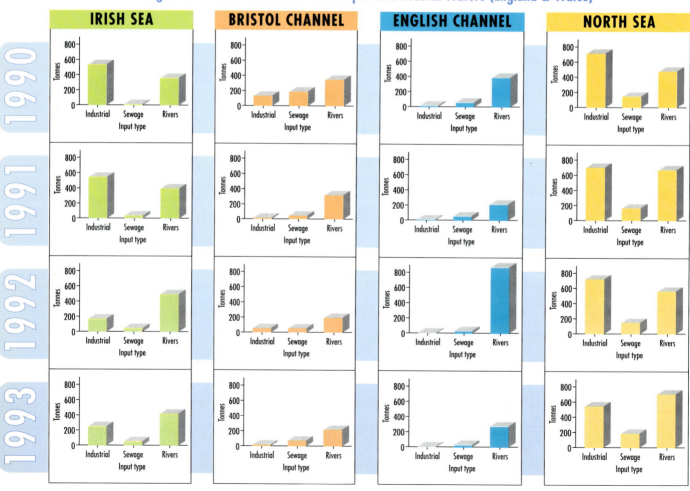

Figure 6.15: Annual Zinc Discharges into Major Estuaries 1990-1993

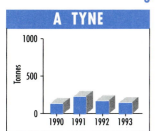

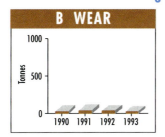

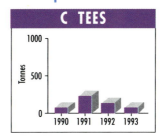

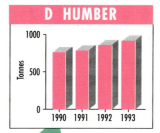

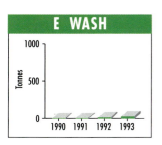

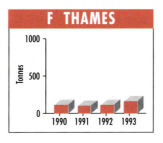

Figure 6.16: Leading Zinc Discharges

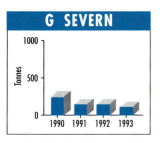

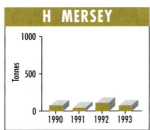

Table 6.5: Zinc Discharges 1990-93, ranked by size as a percentage of the total load measured in each year

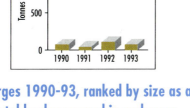

No.	Input Type	NRA Region	Catchment	1990 3118 Rank	%	1991 3089 Rank	%	1992 3275 Rank	%	1993 2610 Rank	%
1	Industrial	Anglian	Humber	1	14.2	1	13.3	2	15.5	1	16.3
2	Industrial	North West	Irish Sea	2	9.2	4	4.3	5	3.9	6	3.6
3	River	South Western	Fal (Restronguet Creek)	3	8.7	6	4.1	1	22.6	3	5.9
4	Industrial	Welsh	Milford Haven	4	5.5						
5	River	Severn Trent	Severn	5	4.3	10	1.9	7	2.3	8	2.6
6	Industrial	South Western	Severn Estuary	6	3.8	9	2.7				
7	Industrial	Welsh	Swansea Bay	7	3.6	2	12.9				
8	River	Northumb/Yorks	Ouse	8	2.3	7	3.9			4	5.7
9	River	Severn Trent	Trent	9	2.2	8	3.4	3	4.4	2	6.0
10	Industrial	Northumb/Yorks	North Sea	10	2.0						
11	River	Northumb/Yorks	Tyne			3	5.2	4	4.1	5	4.3
12	Industrial	Northumb/Yorks	Tees			5	4.3	6	2.8		
13	Industrial	Anglian	Humber					8	1.8		
14	River	Welsh	Ystwyth					9	1.9	9	2.3
15	River	North West	Mersey					10	1.5		
16	Industrial	Welsh	Milford Haven							7	2.9
17	Sewage	Thames	Thames							10	2.3

LINDANE

Concentrations of synthetic organic contaminants, such as lindane (gamma-hexachlorocyclohexane), in environmental samples are usually very low and close to the limit of detection. Therefore there can be large fluctuations in estimated loads of these compounds from year to year because of variations in river flow. Lindane inputs have remained relatively constant over the 1990-1993 period.

Generally the North Sea receives the greatest proportion of lindane, with the Thameside sewage works being the principal sources, although river inputs (probably reflecting upstream sewage and agricultural inputs) are also significant. There is some doubt about the validity of the 1991 Irish Sea data. Subsequent analysis has used more refined techniques and the following years data maybe more robust.

Of the major estuaries in England and Wales, the Thames receives the highest loads of lindane, followed by the Humber. Loads to all estuaries (with the possible exception of the Mersey) are relatively constant within the limits of analytical accuracy over the 1990-1993 period.

Figure 6.17: Annual Lindane Low Load input into Coastal Waters (England & Wales)

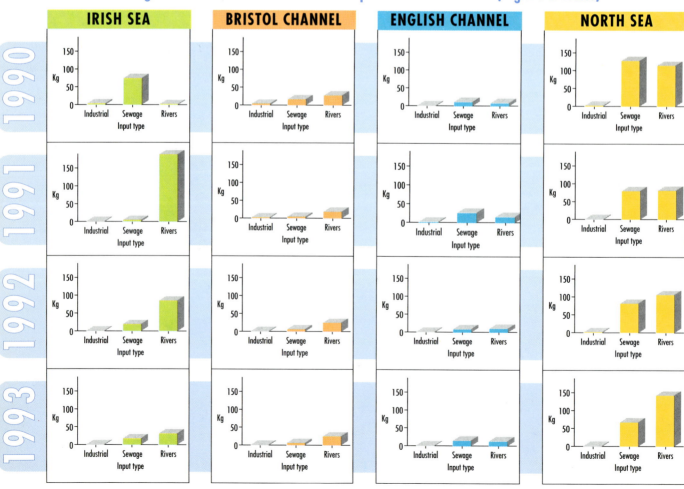

Figure 6.18: Annual Lindane Discharges into Major Estuaries 1990-1993

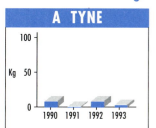

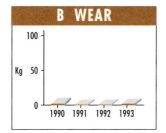

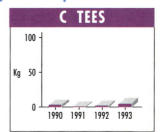

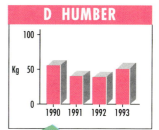

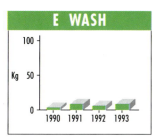

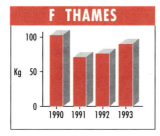

Figure 6.19: Leading Lindane Discharges

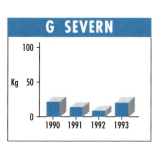

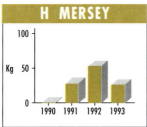

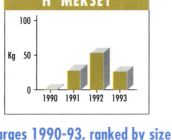

Table 6.6: Lindane Discharges 1990-93, ranked by size as a percentage of the total load measured in each year

No.	Input Type	NRA Region	Catchment	1990 Rank	1990 %	1991 Rank	1991 %	1992 Rank	1992 %	1993 Rank	1993 %
	TOTAL (Kg)			**375**		**407**		**317**		**308**	
1	Sewage	Thames	Thames	1	10.1	3	7.1	2	8.2	3	7.3
2	Sewage	Thames	Thames	2	7.3	6	4.5	8	4.5	7	4.7
3	River	Severn-Trent	Trent	3	7.1	5	4.9	3	6.5	4	6.1
4	Sewage	Welsh	Irish Sea	4	6.1						
5	River	Thames	Thames	5	6.0			5	6.2	1	13.2
6	River	Severn Trent	Severn	6	5.3					5	5.9
7	Sewage	Welsh	Irish Sea	7	4.9						
8	Sewage	Northumb/Yorks	Humber	8	4.4						
9	Sewage	Welsh	Irish Sea	9	3.7						
10	Sewage	Thames	Thames	10	3.7	10	3.5	7	4.8	9	3.4
11	River	Northumb/Yorks	Ouse (Aire)			9	3.7	6	4.8	2	10.3
12	River	North West	Eden			2	9.0				
13	River	North West	Mersey			4	5.3	1	9.9	6	5.3
14	River	North West	Ribble			1	13.6				
15	River	North West	Lune			7	4.2				
16	River	North West	Derwent			8	4.0				
17	River	North West	Weaver					4	6.3		
18	Sewage	Northumb/Yorks	Tyne					9	3.0		
19	Sewage	North West	Ribble					10	2.1		
20	Sewage	North West	Mersey							8	3.4
21	River	Thames	Thames (Lee)							10	3.2

SIMAZINE

Simazine is a herbicide which is widely used for weed control (although restrictions on its non-agricultural use were imposed in 1993). Rain water run-off from treated land is the main source of the simazine entering coastal waters. The concentrations of simazine found in river water and other inputs are low. Thus, load estimates lack precision and can fluctuate significantly from year to year. Care is needed in interpreting the results.

The North Sea and the Irish Sea are the coastal seas which receive the greatest load of simazine, although the loads to both seas have fluctuated through the years. This is likely to be due more to the imprecision with which simazine loads can be estimated than any real variation from year to year. However, overall use of simazine has reduced now that the non-agricultural use is banned and this reduction will gradually be seen in actual loads to sea.

The biggest single source of simazine is the River Trent. The country's other big rivers are also significant sources, as are the large Thameside sewage inputs.

Figure 6.20: Annual Simazine Low Load input into Coastal Waters (England & Wales)

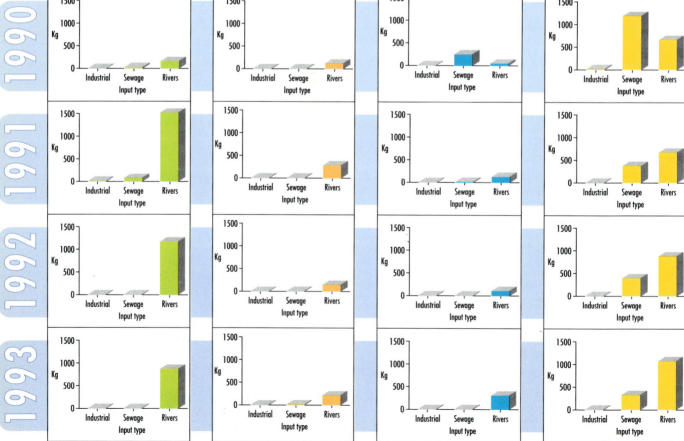

Figure 6.21: Annual Simazine Discharges into Major Estuaries 1990-1993

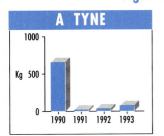

A TYNE

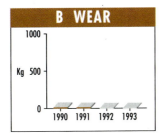

B WEAR

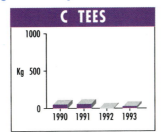

C TEES

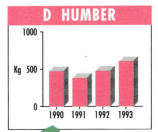

D HUMBER

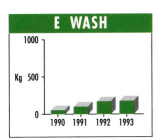

E WASH

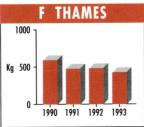

F THAMES

Figure 6.22:
Leading Simazine
Discharges

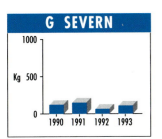

G SEVERN

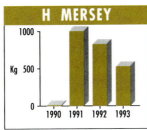

H MERSEY

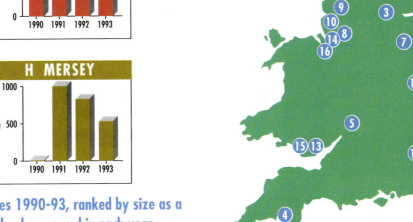

Table 6.7: Simazine Discharges 1990-93, ranked by size as a percentage of the total load measured in each year

No.	Input Type	NRA Region	Catchment	1990 2414 Rank	%	1991 3029 Rank	%	1992 2667 Rank	%	1993 1861 Rank	%
1	Sewage	Northumb/Yorks	Tyne	1	20.3						
2	Sewage	Thames	Thames	2	15.4			4	7.3		
3	River	Northumb/Yorks	Ouse (Aire)	3	11.2			8	3.4	2	12.4
4	Sewage	South West	Plym Estuary	4	9.6						
5	River	Severn Trent	Severn	5	4.6	6	4.4			5	6.2
6	Sewage	Northumb/Yorks	Tyne	6	3.3						
7	River	Severn Trent	Trent	7	2.9	3	6.3	2	8.5	1	21.5
8	River	North West	Mersey	8	2.8	1	24.5	1	25.1		
9	River	North West	Ribble			2	11.5				
10	River	North West	Douglas			4	4.7				
11	River	Thames	Thames			5	4.5			6	5.9
12	Sewage	Thames	Thames			7	4.1	6	5.0	4	9.1
13	River	Welsh	Taff			8	3.9				
14	River	North West	Weaver					5	5.5		
15	River	Welsh	Ogmore					3	8.1		
16	River	Welsh	Dee					7	3.8		
17	River	Southern	Medway							3	10.1
18	River	Northumb/Yorks	Tyne							7	5.3
19	River	Anglian	Great Ouse							8	2.7

TOTAL NITROGEN

Total nitrogen comprises inorganic forms such as ammonia, nitrite and nitrate and also organic forms. The organic forms are usually only a small fraction of the total and are not usually measured by the NRA. Thus, the term total nitrogen used here refers to mean total inorganic nitrogen only.

Nitrogen reaches rivers via leaching and run-off of fertilizers from agricultural land and other diffuse sources. It is a significant component of sewage effluent and can also come from direct discharges of trade effluent from the chemical industry. Run-off from agricultural land is inevitably linked to rainfall, with more rainfall leading to greater leaching.

The largest input of total nitrogen is the River Trent discharging to the Humber, with the River Severn close behind. Details of the other inputs are as shown in Table 6.8 and Figure 6.24.

The North Sea receives the highest loads of nitrogen; over twice that of any other sea area. This reflects the size of the rivers discharging to that sea area. The Humber is the estuary receiving the largest load of nitrogen, as is to be expected given the size of the freshwater river systems which flow to it.

Figure 6.23: Annual Total Nitrogen Low Load input into Coastal Waters (England & Wales)

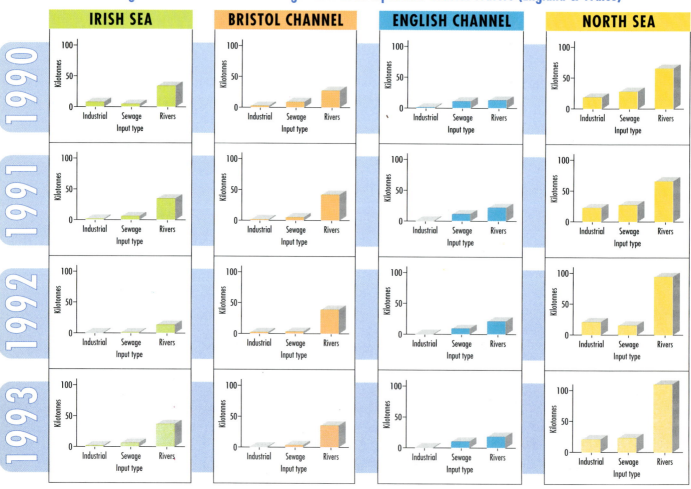

Figure 6.24: Annual Total Nitrogen Discharges into Major Estuaries 1990-1993

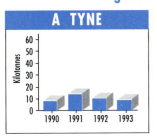

A TYNE

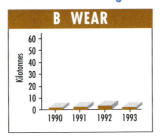

B WEAR

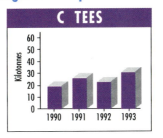

C TEES

D HUMBER

E WASH

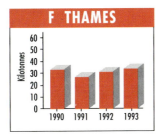

F THAMES

Figure 6.25: Leading Total Nitrogen Discharges

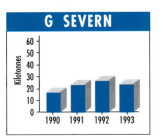

G SEVERN

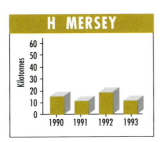

H MERSEY

Table 6.8: Total Nitrogen Discharges 1990-93, ranked by size as a percentage of the total load measured in each year

No.	Input Type	NRA Region	Catchment	1990 211000 Rank	%	1991 230000 Rank	%	1992 179000 Rank	%	1993 274000 Rank	%
1	River	Severn Trent	Trent	1	6.9	2	8.3	1	16.1	1	9.1
2	River	Severn Trent	Severn	2	6.1	1	9.0	2	10.5	2	8.5
3	River	Thames	Thames	3	5.9	6	3.3			3	5.4
4	River	Northumb/Yorks	Ouse	4	4.2			5	3.0	=5	2.8
5	Industrial	Northumb/Yorks	Tees	5	3.9	3	4.4	3	6.9	4	4.8
6	River	North West	Mersey	6	3.2	7	2.3				
7	River	North West	Weaver	7	2.9	10	1.9				
8	Sewage	Thames	Thames	8	2.8	8	2.2				
9	Industrial	Welsh	Milford Haven	9	2.5						
10	River	Northumb/Yorks	Ouse (Aire)	10	2.5	4	3.7	4	3.7	=5	2.8
11	River	Welsh	Wye			5	3.7	6	2.8		
12	Industrial	Northumb/Yorks	Tees			9	2.0				
13	River	Northumb/Yorks	Tweed					7	2.6	10	2.0
14	River	Wessex	Bristol Avon					8	2.5		
15	River	Northumb/Yorks	Ouse (Don)					9	2.2	9	2.0
16	River	South Western	Exe					10	2.0		
17	River	Anglian	Ouse (Ely)							7	2.7
18	River	Anglian	Ouse (Bedford)							8	2.3

ORTHOPHOSPHATE

Phosphorus is present in the aquatic environment in both inorganic and organic forms. However the organic forms contribute only a small proportion of the total and have been disregarded by the NRA in its PARCOM monitoring. The following, therefore, refer to inorganic phosphorus and more particularly to the dissolved orthophosphate form. Orthophosphate, like nitrogen, is a significant component of domestic sewage, but also arises from industrial and agricultural sources.

The single biggest source of phosphate until 1992 was an industrial input on the Cumbrian coast but, following a change in manufacturing practice, the importance of this source has fallen. The Thameside sewage effluents and some large rivers such as the Trent and Severn are the principal sources of phosphate inputs to the sea.

The North Sea receives the highest loads of phosphate, due in part to the size of the rivers which drain into it and, in part, to the size of the conurbations, including Greater London, which produce large sewage effluent loadings.

Figure 6.26: Annual Orthophosphate Low Load input into Coastal Waters (England & Wales)

	IRISH SEA	BRISTOL CHANNEL	ENGLISH CHANNEL	NORTH SEA
1990				
1991				
1992				
1993				

(Each cell contains a bar chart with axis labelled "Tonnes" (0–10000) and x-axis "Input type" showing Industrial, Sewage, Rivers.)

Figure 6.27: Annual Orthophosphate Discharges into Major Estuaries 1990-1993

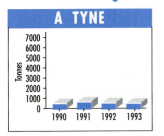
A TYNE

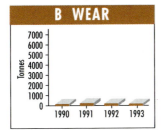
B WEAR

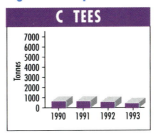
C TEES

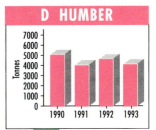
D HUMBER

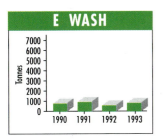
E WASH

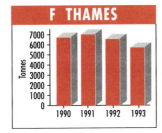
F THAMES

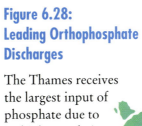

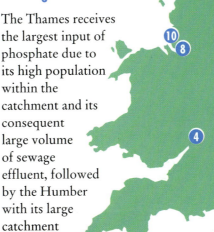

Figure 6.28: Leading Orthophosphate Discharges

The Thames receives the largest input of phosphate due to its high population within the catchment and its consequent large volume of sewage effluent, followed by the Humber with its large catchment area (25% of England) and the number of large conurbations within it account for this high load.

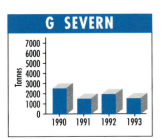
G SEVERN

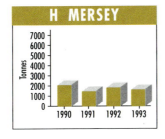
H MERSEY

Table 6.9: Orthophosphate Discharges 1990-93, ranked by size as a percentage of the total load measured in each year

				1990 31,600		1991 30,400		1992 27,500		1993 23,500	
No.	Input Type	NRA Region	Catchment	Rank	%	Rank	%	Rank	%	Rank	%
1	Industrial	North West	Irish Sea	1	12.4	1	19.5	2	9.4		
2	Sewage	Thames	Thames	2	9.6	3	6.9	3	6.0	3	7.2
3	River	Severn Trent	Trent	3	8.6	2	8.0	1	10.8	1	13.0
4	River	Severn Trent	Severn	4	5.9	5	3.7	5	5.0	2	7.4
5	Sewage	Thames	Thames	5	4.7	4	3.9	8	2.8	6	3.8
6	Sewage	Thames	Thames	6	3.3	6	3.6	4	5.8	5	4.1
7	River	Thames	Thames	7	3.1	7	3.3	6	3.5	4	6.7
8	River	North West	Mersey	8	2.9	8	2.7	7	3.4		
9	River	Northumb/Yorks	Ouse (Aire)	9	2.0	10	1.8	10	2.0	7	3.4
10	Sewage	North West	Mersey	10	1.7						
11	Sewage	Northumb/Yorks	Tyne			9	2.0				
12	River	Thames	Thames (Lee)					9	2.2	8	2.8
13	Sewage	Southern	Medway							9	2.2
14	River	Northumb/Yorks	Ouse (Don)							10	2.1

43

Chapter 7   EUROPEAN PERSPECTIVE

At the Second North Sea Conference in London in 1987, Ministers agreed to prepare national priority lists of substances whose input to the North Sea would be reduced. With the exception of France, each country presented a national priority list at the Third North Sea Conference, held at the Hague in 1990. There was general agreement on which substances were considered to pose the greatest threat and this consensus was used to prepare a Common Priority Action List of 36 substances (also known as the Annex 1A List) with target reductions. However, each country was allowed to select its own method of achieving load reductions.

The actions taken by signatory countries can broadly be broken down into: the listing of priority substances; identifying the means by which inputs from point and diffuse sources could be reduced; estimating baseline figures against which reductions in inputs can be gauged; and identification of suitable monitoring systems to record input reductions.

7.1 NATIONAL LISTS OF PRIORITY HAZARDOUS SUBSTANCES

All signatory countries drew up national lists prior to the 1990 Conference. Most of these have been modified to ensure that they are compatible with the Common Priority List. In general, countries have concentrated on reducing inputs of substances from the Common Priority List that were on their own National Lists.

Prior to the production of the Red List in 1989 the UK did not have a set national list. Instead a priority candidate list was used informally to identify the potentially most polluting substances.

7.2 TECHNICAL APPROACHES TO INPUT REDUCTION

As has been described earlier in the report inputs of contaminants to the aquatic environment may arise from two sources:

- **point sources such as industrial or sewage effluents;**
- **diffuse sources such as agricultural activities.**

To date all countries have concentrated on reducing inputs from point sources, largely because control mechanisms are more straightforward and easier to enforce.

7.2.1 POINT SOURCES

Two main control methods are available for reducing inputs from point sources:

- **end of pipe control;**
- **clean technology.**

The application of end of pipe control requires the treatment of waste to reduce concentrations in effluents discharged to the aquatic environment and is a process commonly referred to as Best Available Techniques (BAT).

Most countries are also introducing clean technology which aims to reduce the environmental impact of industrial effluents by avoiding the production of dangerous waste. This can be achieved by a variety of measures including altering production processes and using less hazardous substances in product formulations.

In the UK Integrated Pollution Control (IPC), introduced in 1990, uses Best Available Techniques Not Entailing Excessive Cost (BATNEEC) to control the input of 'prescribed' substances to the environment from certain industrial processes; for water these prescribed substances are the Red List.

A new directive on Integrated Pollution Prevention and Control is currently under discussion at the EC. This is similar in concept to the UK system of IPC where all the discharges and emissions from a process are considered before an authorisation is given for discharge to the environment. The Directive also embraces the concept of BAT, which encompasses the requirement that benefits and costs must be taken into account.

The other European countries use concepts broadly similar to the UK's BATNEEC. Some countries such as Germany use water pollution levies over and above regulatory requirements to encourage dischargers to reduce inputs of hazardous substances. In Scandinavian countries the emphasis is on the phasing out of the use of most hazardous chemicals.

In all countries it is likely that the reduction of inputs from municipal sewage works will be sought at source by applying BAT to the discharge of hazardous substances into the sewer rather than installing extra treatment facilities at the treatment works.

7.2.2 DIFFUSE SOURCES

Inputs from diffuse sources are more difficult to identify and control compared to those from point sources. Controls available include:

- **banning the use of substances altogether;**
- **stipulating or restricting methods of using hazardous substances;**
- **changing product formulations to minimise the presence of hazardous substances.**

No country has yet introduced comprehensive codes of practice to control diffuse inputs but some significant initiatives have been taken.

Denmark, Germany, the Netherlands, Sweden and the UK have all used legislative measures to reduce the use of some of the Common List substances. Norway has imposed environmental taxes on certain priority substances.

7.3 SELECTION OF BASELINE FIGURES

In order to gauge the reduction of inputs of substances on the Common Priority List all signatory countries need to establish a baseline for 1985 discharges.

Most countries have concentrated on estimating baseline figures for pollutants that appeared on their own National Lists prior to 1990. Where possible, 1985 has been used as the base year, however, where this is not possible other years have been used, or computer models used to extrapolate from more recent figures.

So far the UK has been able to provide 1985 baseline figures for the metals mercury, cadmium, copper, zinc, lead, chromium and nickel as well as lindane. Work is continuing to improve the estimates of baseline figures for other substances.

More comprehensive data have been provided by Germany and the Netherlands which have calculated inputs of many of the individual organic substances on the list as well as inputs of metals from industrial, municipal and diffuse sources. In Scandinavia substances were originally grouped together on priority lists, hence baseline data tends to relate to groups of substances. France did not present estimated input levels for 1985 at the Third North Sea Conference but, for Germany, input levels to the Rhine have been calculated from an inventory of dangerous substances made in 1988 using data obtained for 1985.

All countries have had difficulties in establishing baseline figures for diffuse sources.

7.4 MONITORING REQUIREMENTS

To quantify the reductions in inputs achieved, river monitoring and/or monitoring of point sources may be used. Most countries prefer to rely on point source monitoring, sometimes with some river monitoring, but the UK uses river monitoring as a primary means of gathering data. Germany and Belgium have introduced monitoring systems at the tidal limits of the Weser, Elbe and Schelde and on the Rhine at the Dutch/German border.

7.5 UK ACTIONS COMPARED TO OTHER COUNTRIES

For purposes of estimating inputs and reporting to the Fourth North Sea Conference, the UK monitors the quantity of priority substances entering the North Sea from each of its major rivers at the tidal limit. In addition, the UK takes into account direct discharges to estuaries and coastal waters downstream of the tidal limit. This provides a direct estimate of all UK inputs to the North Sea. Other signatory countries rely on monitoring point sources and, in some cases, estimating inputs from diffuse sources to give an indirect estimate of load discharged to the North Sea.

7.6 UK RESULTS COMPARED TO OTHER COUNTRIES

A direct comparison of the estimated results from all the signatory countries is not easy because of the basic differences in methods of estimating inputs. The UK approach gives estimates that reflect actual inputs to the North Sea. Other countries provide estimates which identify reductions at sources more specifically. It is possible to assess the performance of each country in achieving reductions within its own estimating method.

Results of the measurement programmes will be published for the Fourth North Sea Conference.

Chapter 8

NRA ACTIONS AND FUTURE DEVELOPMENTS

For both the PARCOM and the North Sea Conference surveys, and for both high and low load data on dangerous substances, the importance of individual inputs is assessed by the NRA. The percentage that each input contributes to the national total load is calculated and inputs are then ranked in order of size.

Such ranking has to be interpreted with some care. In those cases where an individual load is based on ten or more positive results and on ten or more measured flow recorder results, the accuracy of the data is relatively high. However, where a more limited number of the samples analysed through the year have produced positive values, or where flow has had to be estimated rather than measured, then the accuracy of the load calculation is lower.

In spite of the caution required when interpreting the data, the NRA believes ranking individual loads may be a useful tool for identifying the major sources of input of individual substances and is investigating further.

8.2 REVIEWING CONSENTS

Before any effluent can be discharged, whether industrial or sewage, to a river, estuary or direct to sea it needs the written consent of the NRA. In granting the consent the NRA imposes conditions on the volume and the nature and composition of the effluent to control the load that can be discharged. Where dangerous substances are known to be present in the effluent in significant amounts, the consent will normally identify those substances and impose a concentration limit and a restriction on volume. Consequently, a limit on the maximum load that can be discharged is imposed on the discharger.

Where a specific catchment is found to be a major source of a particular substance then discharge consents containing that substance will be reviewed. While the individual cases will each have to be judged on their merits, the overall objective is to reduce the permitted load that can be discharged from the catchment.

The introduction of Integrated Pollution Control (IPC) means that some discharges containing dangerous substances will be authorised under the Environmental Protection Act 1990, by Her Majesty's Inspectorate of Pollution (HMIP). In these cases the condition on the authorisation will be set at levels at least as stringent as those required by the NRA. In formulating the NRA requirements, the same approach will be adopted, ie. an agreed reduction in the load discharged will be set.

8.3 SURVEYS OF FRESHWATER CATCHMENTS

Chapter 6 highlighted the fact that some freshwater rivers are a major source of certain substances. In such cases, an assessment is required to determine whether the presence of the substance in the river is a result of point or diffuse inputs. This involves the identification of key sources of the substance in the catchment and an evaluation of their contributions to the river system.

Typically, such an assessment will commence with an examination of the consents for effluent discharges within the catchment. This will identify any point sources of the substance under investigation. Indeed in some cases this initial examination of the available data will indicate that most of the substance load is introduced by effluents and a review of the discharge consents will be required to achieve the necessary load reductions.

However, if it is clear from a review of the upstream discharges, that the major source of a substance is from diffuse inputs, it becomes difficult for the NRA to achieve load reductions. The methods normally employed for reducing diffuse inputs of substances, such as pesticides, are to alter the way in which the substance is used, to encourage a replacement with a more "environmentally friendly" version, or to press for a complete ban of a particular substance.

Where diffuse inputs are the major source of a substance in a river, the NRA tries to ensure that river monitoring is comprehensive enough to identify particular areas of the catchment which appear to be the main sources of contamination. Control mechanisms available to the NRA will almost certainly need to involve Central Government (eg. MAFF) in any initiative to reduce or change the use of a particular substance, such as a pesticide. Comprehensive monitoring will need to be continued on the upstream freshwater river for some years to confirm that loads have been reduced.

8.4 CONTROL OF NUTRIENTS

The NRA is involved in the implementation of two EC Directives concerning the control of nutrient inputs to surface waters. The Urban Waste Water Treatment Directive (91/271/EEC) addresses the issue of pollution from point source discharges, particularly sewage treatment works (STWs). In contrast, the Nitrate Directive (91/676/EEC) is intended to reduce water pollution by nitrates from agricultural (primarily diffuse) sources. Both Directives are in the early stages of implementation and the nutrient control measures will begin to come into force over the next 3-5 years.

Under the Urban Waste Water Treatment Directive, discharges from large STWs into waters identified as Sensitive Areas may require nutrient reduction treatment. Sensitive Areas may be defined in relation to waters affected by eutrophication, where high concentrations of either phosphorus or nitrogen give rise to enhanced plant growth, leading to undesirable effects upon the water quality and the organisms present. On the advice of the NRA, the Government has identified 33 freshwater sites in England and Wales as Sensitive Areas (eutrophic). As a consequence, some 42 large STWs will require phosphorus reduction treatment by the end of 1998. No waters were found to be eutrophic as a result of elevated nitrates. Designations must be reviewed at least every four years. The NRA is undertaking monitoring of both fresh and saline waters, to provide data for use in assessing whether further sites should be designated in the first review (1997). A further category of Sensitive Area may be identified, relating to waters where large STWs contribute to high nitrate concentrations at public drinking water abstraction points. The UK Government has not as yet announced any designations in this category.

Under the Nitrate Directive, areas of land draining to waters affected by pollution, from agricultural nitrates, must be designated as Vulnerable Zones. The relevant waters are those found to be eutrophic due to nitrates, those with high nitrate concentrations at public drinking water abstraction points, together with high-nitrate groundwaters. The Government proposes to designate the land draining parts of 11 river systems used for drinking water abstraction plus that draining to some 140 groundwater abstractions, a total area of some 650,000 hectares in England and Wales. As part of this process the NRA provided advice and data to the Government on the quality of the waters and the movement of water. The NRA will continue to monitor affected waters to provide information for four-yearly designation reviews. Within the Vulnerable Zones, once designated, action programmes will be introduced, to control agricultural practices which give rise to nitrate leaching.

In addition to this European legislation, there are domestic initiatives aimed at reducing pollution of water by nutrients. The NRA has assisted the Government (MAFF) in defining Nitrate Sensitive Areas (NSAs), being areas where agricultural activities are restricted to reduce water pollution by nitrates. This initiative is aimed at improving the quality of groundwater sources. However, surface water systems within the NSAs will also benefit from the implementation of the measures. The NRA is also involved in promoting good practice within the agricultural community, aimed at preventing water pollution in general (including nutrient pollution). This is achieved though a range of measures including regulations (for example those relating to the construction of silage and slurry stores) and educational initiatives. The NRA contributed to the MAFF Code of Good Agricultural Practice for the Protection of Water which seeks to promote best practice and prevent pollution.

There are 18 pesticides on the Annex 1A list. With the exception of Pentachlorophenol (PCP) and some of the organochlorines, the majority of the load derived from these compounds comes from diffuse run-off; ie. from the land rather than discharges from industry or domestic sources. This diffuse pollution can result either from the use of pesticides to protect agricultural crops or from weed control in non-agricultural situations.

Diffuse source run-off of pesticides is difficult to control. However it is sometimes possible to withdraw a pesticide from use in situations where the run-off is significant; this method has been adopted for the organochlorines and for the non agricultural use of the triazine herbicides, atrazine and simazine. Alternatively steps can be taken to reduce the overall use of the pesticide. And even where there is no reduction in use, it may be possible to improve the way they are used so that run off is minimised.

The NRA's strategy for reducing pesticide pollution includes measures to control point source discharges directly by using powers under the Water Resources Act (1991) and to influence pesticide use patterns to reduce pollution from diffuse sources. Where the NRA's monitoring programmes indicate that pesticides are present in the aquatic environment in significant quantities the evidence is passed to the relevant Government Departments. They can then address the problem during reviews of approval for use. These reviews are carried out under the Food and Environment Protection Act (1985) at present but in future it will be through implementation of the EC "Uniform Principles" Directive. These reviews may result in the restriction of use of the pesticide or the selective withdrawal of approval under those situations where it is known to be causing a particular problem or, in extreme cases, the complete withdrawal of the approval for use.

Of equal importance in the NRA's strategy is the encouragement of Best Practice in the manufacture, transport, storage, use and disposal of pesticides. This has included the significant inputs to statutory and non-statutory guidance on the use, storage and disposal of pesticides, particularly in relation to the measures included for pollution prevention. This aspect is seen to be particularly important in reducing the number of small incidents resulting from use and disposal. In addition, the NRA has produced its own guidance leaflet for farmers to increase awareness of the potential problems which can be caused by pesticides in water.

GLOSSARY OF TERMS

BIOACCUMULATION The mechanism whereby organisms concentrate in their tissues heavy metals, or other stable compounds, present in dilute concentrations in saline or fresh water.

BIOCIDE Any substance that kills biological life eg. pesticides, insecticides, fungicides etc.

CONGENERS The different chemical forms that a single generic substance can take.

CONSENT A statutory document issued by the NRA to indicate any limits and conditions on the discharge of an effluent to a controlled water.

CONTROLLED WATERS All rivers, canals, lakes, groundwaters, estuaries and coastal waters to three nautical miles from the shore.

DANGEROUS SUBSTANCES Substances defined by the European Commission as in need of special control because of their toxicity, bioaccumulation and persistence. The substances are classified as List I or List II according to the EC Dangerous Substances Directive (74/464/EEC).

DETERMINAND Literally 'that which is to be determined'. A general term for any numerical property of a sample (eg. the amount of cadmium) whose value is required.

DIRECTIVE A type of legislation issued by the European Community which is binding on Member States in terms of the results to be achieved.

ECOSYSTEM A biological community of interacting organisms and the physical environment associated with it.

ENVIRONMENTAL QUALITY OBJECTIVE (EQO) The requirement that a body of water should be suitable for certain identified uses.

ENVIRONMENTAL QUALITY STANDARD (EQS) A specific concentration limit for a particular substance which affects a particular water use or objective.

EUTROPHICATION The natural ageing of a lake or land-locked body of water which results in organic material being produced in abundance due to a ready supply of nutrients accumulated over the years. This process can be rapidly increased by man due to nutrients from agriculture and sewage treatment processes.

HALOGENATION Reaction with a member of the halogen group : ie. fluorine, chlorine, bromine, iodine.

HEAVY METALS A group of metals with high atomic mass, which pose specific environmental problems due to their toxicity, persistence, and tendency to persist in living systems (bioaccumulate).

LEACHING The removal by water of any soluble constituent from the soil.

LIMIT OF DETECTION (LOD) The lowest concentration of a determinand that can be successfully detected by a given analytical method. This can be calculated statistically.

LIMIT VALUES A method of fixing the amount of waste substance which may be present in an input according to the manufacturing process from which it arises rather than the effect it may have on the receiving environment. Commonly expressed as kilograms of waste permitted per tonne of manufactured product.

NON-AROMATIC An organic compound that does not contain a benzene ring and does not have chemical properties similar to benzene.

NUTRIENTS The food of specific organisms. Commonly understood to mean plant nutrients, such as nitrogen and phosphorus, in water.

PARIS COMMISSION The international body that oversees the implementation of the Paris Convention.

PARIS CONVENTION The Paris Convention for the Prevention of Pollution from Land-Based Sources. A response by states bordering the North Sea to reduce inputs of toxic substances and nutrients into the sea via rivers and estuaries.

PERSISTENCE Ability to continue in existence.

PRECAUTIONARY APPROACH The taking of action even when there is no conclusive evidence of a cause and effect relationship between input and an adverse occurrence.

RIPARIAN Land which is in contact with water during ordinary high tides eg. a coastline.

TOXICITY A characteristic of a substance defining its poisoning effect on an organism.

CHEMICAL CHARACTERISTICS OF SUBSTANCES

ALDRIN

CAS NUMBER	309-00-2
CHEMICAL NAME	[1R,4S,4aS, 5S,8R, 8aR]-1,2,3, 4,10,10-hexachloro-1,4,4a,5,8, 8a-hexahydro 1,4:5,8-dimethano-naphthalene
SYNONYMS	aldrex; aldrite; octalene; aldron; aldrosol; algran; altox; compound 118; drinox
ISOMERS	Isodrin
FORMULA	$C_{12} H_8 Cl_6$

USE   Aldrin is an organochlorine insecticide and has been banned in the UK since 1989.

TOXICITY   Aldrin is of high toxicity to aquatic organisms.

Fish: Freshwater   96h LC50, *Lepomis macrochirus,* (bluegill sunfish), 0.013 mg/l
96h LC50, *Oncorhynchus mykiss,* (rainbow trout), 0.036 mg/l
96h LC50, *Lebistes reticulatus,* (guppy), 0.020 mg/l
96h LC50, *Cyprinus carpio,* (carp) 0.004 mg/l (Verschueren, 1983)

Marine   96h LC50, *Roccus saxatilis,* (striped bass), 0.010 mg/l

Invertebrates: Freshwater   24h LC50, *Daphnia sp.* 0.030mg/l
48h LC50, *Daphnia sp.* 0.028mg/l (WHO, 1989)

Algae: Marine   Aldrin has been shown to cause an 85% decrease in the productivity of a marine phytoplankton (Butijin and Koeman, 1977).

BIOACCUMULATION   Aldrin has been shown to bioaccumulate with BCFs (Biological Concentration Factors) in the region of tens of thousands and significant food chain biomagnification has been observed.

ENVIRONMENTAL FATE AND PERSISTENCE   Aldrin is very persistent. It inhibits nitrification and is slowly converted to dieldrin in water (Mailhot and Peters, 1988). Aldrin is insoluble in water and occurs only at low concentrations in the aquatic environment but, because it is strongly adsorbed to soils, it may be associated with sediments.

STANDARDS   Aldrin is a List 1 substance under the Dangerous Substance Directive (76/464/EEC) and a UK Red List substance. The EC have set a statutory Environmental Quality Standard of 10ng/l for all waters. Standards for aldrin may also be included in other legislation.

ARSENIC

CAS NUMBER   7440-38-2

FORMULA   As

USE   Arsenic is used as a wood preservative, and in the manufacture of glass, alloys, medicines and semiconductors. It is also a by-product in the smelting and titanium dioxide industries. UK use is estimated at less than 1000 t/y (Mance *et al* 1984).

TOXICITY

Fish: Arsenic is of moderate toxicity to fish and invertebrates.

    Freshwater   96h LC50, *Oncorhynchus mykiss,* (rainbow trout), 10.8mg/l (Penrose, 1974).

    Marine   96hLC50, *Menidia menidia* (Atlantic siverside), 15 mg/l (As) (US EPA, 1980)

Invertebrates: Freshwater   48h EC50, (immobilisation), *Daphnia magna*, 9.1 mg/l ($NaAsO_2$), (Sanders, 1979)
48h EC 50 (mortality and immobility), *Daphnia pulex,* 1.27 mg/l (Mance *et al* 1984)

BIOACCUMULATION   Arsenic has a low potential to bioaccumulate. BCFs (Biological Concentration Factors) in the range of 1.4-330 have been reported for *Chlorella vulgaris* and <1 for *Oncorhynchus mykiss* (rainbow trout) (Ferguson and Gavis, 1972). Bioaccumulation in marine species may be greater than in freshwater species with a BCF value for mussels *(Mytilis edulis)* ranging from 360-4730. Arsenic has not been shown to biomagnify in food chains, moreover a large proportion of the accumulated arsenic may be present as arsenobetaine posing little threat to the organism (Mance *et al* 1984).

ENVIRONMENTAL FATE AND PERSISTENCE   In water arsenic is usually found in the form of arsenate or arsenite. Arsenic forms covalent bonds with most non-metals and some metals. Usually in aquatic systems the inorganic form will predominate. This can be adsorbed onto hydrous iron and aluminium oxides or can be adsorbed by suspended matter which remove arsenic to sediments. Aquatic organisms are able to transform, accumulate and transport arsenic. Biologically induced transformations can result in the formation of dimethyl arsine and methyl arsonic acid (Mance *et al* 1984).

STANDARDS   Arsenic is a List 2 substance under the Dangerous Substance Directive (76/464/EEC). The DoE have set an Environmental Quality Standard (EQS) of 50ug/l (dissolved, Annual Average) for the freshwater environment and 25ug/l (dissolved, Annual Average) for the marine environment. Standards for arsenic may also be included in other legislation.

ATRAZINE

CAS NUMBER 1912-24-9

CHEMICAL NAME 2-chloro-4-ethylamino-6-isopropyl-amino-1,3,5-triazine

SYNONYMS 6-chloro-N2-ethyl-N4-isopropyl-1,3,5-triazine-4-diamine

FORMULA $C_8 H_{14} Cl N_5$

USE Atrazine is a herbicide. It is permitted for use in agriculture but in September 1993 was banned for use in non-agricultural situations. Production in the UK has been discontinued, since 1993, although there are several formulation and distribution plants in the UK. EC production and usage are estimated at around 17000t/y and 6000t/y respectively (Henriet et al., 1989). Atrazine is often found to exceed the 0.1ug/l limit for pesticides in drinking water in the UK.

TOXICITY Atrazine is of moderate toxicity to aquatic organisms

Fish: Freshwater 96h LC50 *Lebistes reticulatus* (guppy), 4.3 mg/l
96h LC50 *Oncorhynchus mykiss* (rainbow trout), 4.5 - 8.8 mg/l, (The Pesticide Manual, 1987)

Invertebrates: Freshwater 48h EC50, *Daphnia pulex,* 36.5 mg/l

Marine 96h LC50, *Crassostrea virginica* (eastern oyster), 730 mg/l, (Hartman and Martin, 1985)

Algae: EC50 (inhibition of photosynthesis) phytoplankton species, 0.1 - 0.5 mg/l (Moorhead and Koswinski, 1986)

BIOACCUMULATION Atrazine does not show a tendency to bioaccumulate. A BCF of 0.9 has been reported for *Pimephales promelas* (fathead minnow) (Howard, 1990).

ENVIRONMENTAL FATE AND PERSISTENCE Atrazine is moderately mobile in soil (leached 0.5m in 40 days) and stable in neutral media; it has a half life of 60 days in soil. It is persistent in the aqueous environment with a reported half life typically in the range of months (Helling et al., 1988). The major removal process is through degradation within the soil, (Makay et al., 1985). In aquatic systems the main routes of removal are photo-enhanced hydrolysis to 2-hydroxy derivatives. A half-life of 244 days (25°C, pH4) has been reported (Howard, 1990) Partitioning will depend on the water conditions. Diffuse inputs of atrazine into the aquatic environment arise from surface run-off.

STANDARDS Atrazine is a Red List substance and the DoE have derived an Environmental Quality Standard of 2ug/l (Annual Average) and 10ug/l (Maximum Allowable Concentration) in both the freshwater and marine environments. Standards for atrazine may also be included in other legislation.

AZINPHOS-ETHYL

CAS NUMBER	2642-71-9
CHEMICAL NAME	s-(3,4-dihydro-4-oxobenzo [d]-[1,2,3]- triazine-3-ylmethyl) O,O-diethyl-phosphorodithioate

SYNONYMS	Bayer 16259, ethyl gustathion, ethyl guthion, gustathion A, gustathion ethyl, gustathion H and K, guthion ethyl.
FORMULA	$C_{12} H_{16} N_3 O_3 P S_2$
USE	Azinphos-ethyl is used exclusively in agriculture as an insecticide, but was withdrawn in the UK in the 1980's (Worthing, 1991). Annual production in the EC is 1000-2000 t/y.

TOXICITY

Fish:	Freshwater	*Carassius auratus* (goldfish), 96h LC50, 0.1 mg/l
Invertebrates:	Freshwater	48h EC50 (immobilisation), *Daphnia magna*, 0.2 ug/l
Algae:	Freshwater	96h EC50 (growth inhibition), *Scenedesmus subspicatus*, 3 mg/l/l

BIOACCUMULATION	It can be expected to bioaccumulate in some organisms.
ENVIRONMENTAL FATE AND PERSISTENCE	Chemical degradation is the main breakdown pathway. High adsorption onto soil particles and rapid degradation occurs within a few days. The half life at pH 4, 7 and 9 is 3 years, 270 days, and 11 days respectively. Leaching from soils is unlikely to occur and generally azinphos ethyl is not detected in surface waters.
STANDARDS	An EQS for Azinphos-ethyl has not been derived in the UK.

AZINPHOS-METHYL

CAS NUMBER

86-50-0

CHEMICAL NAME

S-(3,4-dihydro-4-oxobenzo
(d)-(1,2,3)-triazin-3-ylmethyl)
O,O-dimethyl phosphoro-
dithioate

SYNONYMS

azinphos-methyl, methyl
azinphos, metriltriazotion, Azinugec, Azimil,
Gusathion, Gusthion methyl, Guthion, methylguthion,
Paricide, Sepizin, Toxation, Crysthion, Carfene

FORMULA

$C_{10} H_{12} N_3 O_3 P S_2$

USE

Azinphos-methyl is a broad spectrum organophosphorous pesticide which was used to treat aphids, weevils and moths and also pests of apples, pears, black currants. Production in the EC was 1500t/yr in 1988. It was withdrawn in the UK in the early 1990s.

TOXICITY

High acute and chronic toxicity to marine and freshwater organisms.

Fish: Freshwater

96h LC50, *Oncorhynchus mykiss* (rainbow trout), 14ug/l (Verschueren, 1983)

Invertebrate: Freshwater

24h LC50 *Gammarus fasciatus*, (shrimp), 0.5 ug/l 48h EC50 *Daphnia magna*, 1.6ug/l (Mayer and Ellersieck, 1986)

BIOACCUMULATION

Azinphos-methyl does not bioaccumulate to any great extent and has a calculated BCF of 72 (Howard, 1990).

ENVIRONMENTAL FATE AND PERSISTENCE

Azinphos-methyl has a low solubility in water. The half life of azinphos-methyl in aqueous solution is 2-28 days, in soil is months, and once absorbed to sediments can remain for up to 17 years. Azinphos-methyl is broken down under UV light and is also biodegraded in activated sludge but only at low concentrations. Contamination of surface waters occurs mainly from diffuse sources. Adsorption to sediments is likely to be a significant removal process. Azinphos-methyl undergoes oxidation to the phosphate (oxon) and thiol phosphate; demethylation of one or both methyl groups and hydrolysis also occurs to give phosphoric acid. Biodegradation will probably be the most important removal process. (Howard, 1990) The degradation products are of a lower toxicity than the parent compound.

STANDARDS

Azinphos-methyl is a Red List substance and the DoE have derived an EQS for both the freshwater and marine environments of 0.01ug/l (Annual Average) and 0.04ug/l (Maxiumum Allowable Concentration). Standards for azinphos-methyl may also be included in other legislation.

CADMIUM

CAS NUMBER   7440-43-9

FORMULA   Cd

USE   Cadmium is used in the production of alloys, solders, deoxidisers, electrodes, photoelectric cells, dental amalgam, and as antihelminthics for swine and poultry.

TOXICITY   Cadmium is of high toxicity to freshwater organisms. Marine species are less sensitive to cadmium with toxicities being moderate-high. Cadmium is also toxic to a wide range of micro-organisms. (WHO, 1992).

Fish: Freshwater   96h LC50,*Oncorhynchus mykiss* (rainbow trout) 1.3-2.6 mg/l
Salmo salar (Atlantic salmon) acute toxicity threshold 156 ug/l

Invertebrates: Freshwater   *Daphnia magna* acute toxicity threshold 0.15 ug/l

    Marine   48h LC50, *Argopecten irradians* (bay scallop), 3.21mg/l
96h LC50, *Eurytemorra affinis* (copepod), 0.06 mg/l (WHO, 1992)

BIOACCUMULATION   Cadmium has been found to bioaccumulate in some organisms with freshwater BCF values range from 164-4190 for invertebrates and 3-2213 for fish. The corresponding values for marine invertebrates and fish are >1000 and 5-3160. Inorganic cadmium complexes are not taken up by fish. Cadmium is translocated by aquatic plants and concentrated in roots and leaves. It is also taken up by aquatic organisms. A BCF of 15 (wet weight) and 200 (ash weight) has been reported for *Fundulus heteroclitus* (mummichog) and a BCF of 190 has been recorded for *Pimephales promelas* (fathead minnow) (WHO, 1992).

ENVIRONMENTAL FATE AND PERSISTENCE   Cadmium is strongly adsorbed to sediments. The free metal ion Cd^{2+} is the form of cadmium most readily available to aquatic organisms.

STANDARDS   Cadmium is a List 1 substance under the Dangerous Sustance Directive (76/464/EEC). The EC have set a statutory EQS of 5ug/l (Total) in freshwater and 2.5ug/l (Dissolved) in the marine environment. Standards for cadmium may also be included in other legislation.

CHLOROFORM

CAS NUMBER	67-66-3
CHEMICAL NAME	Trichloromethane
SYNONYMS	Formyl trichloride, methane trichloride, methyl chloride, methyl trichloride, trichloroform, TCM, Freon 20, R20
FORMULA	$CH Cl_3$

USE

Chloroform is an industrial solvent of moderate solubility in water; the ocean is a major source of natural chloroform. Chloroform is used in the UK in the synthesis of chlorodifluoromethane and as a solvent, fumigant, and in the manufacture of anaesthetics. Production in the EC was estimated as 50 000 t/y in 1988.

TOXICITY

Chloroform is of low-moderate acute toxicity to aquatic organisms but extremely toxic to anaerobic organisms.

Fish: Freshwater

96h LC50, *Oncorhynchus mykiss* (rainbow trout), 43.8-66.8 mg/l (USEPA, 1980). The acute toxicity threshold for both *Oncorhynchus mykiss* (rainbow trout) and *Lepomis macrochirus* (bluegill sunfish) is 18 mg/l

Invertebrates: Freshwater

48h LC50, *Daphnia,* 29 mg/l (Leblanc, 1980)

Marine

48hr LC50, larval oyster, 1mg/l

BIOACCUMULATION

There is no evidence for significant biomagnification up the food chain.

ENVIRONMENTAL FATE AND PERSISTENCE

Chloroform is not persistent, due to its volatility, with a half life of 18.2 - 25.7 minutes. Sources of chloroform in water include discharges from the organic chemistry industry and from chlorination of water supplies. The major route of removal from water bodies is through volatilisation (Dilling *et al,* 1975). Chloroform is reported to have a volatilisation half-life of 36 hours in river water, 40 hours in pond water and 9-10 days in lake water (Howard, 1990).

STANDARDS

Chloroform is a List 1 substance under the Dangerous Substance Directive (76/464/EEC). The EC have set a statutory EQS of 12ug/l for all waters. Standards for chloroform may also be included in other legislation.

CHROMIUM

CAS NUMBER   7440-47-3

FORMULA   Cr

USE   Chromium is used in iron and steel production, metal plating, in pigment production, during the manufacture of glass, in the leather tanning industry and in lithographic and photographic applications (WHO, 1988). World production in 1985 was 11 million tonnes.

TOXICITY   Chromium III and VI differ in toxicity, but with no clear trends. To some organisms and under certain conditions chromium VI is more toxic, but in other cases chromium III may be more toxic.

Fish: **Freshwater**   The lowest adverse effect concentration recorded was a 70% mortality of *Salmo salar* (Atlantic salmon) following exposure to 0.1 mg/l chromium VI for 113 days.
96h LC50, *Oncorhynchus mykiss*, (rainbow trout), 3.4-69 mg/l (hexavalent chromium) (Mance *et al.*, 1984c)

Marine   96h LC50, *Limanda limanda*, (dab) and *Chelon labrosus* (grey mullet), 47mg/l (hexavalent chromium)

Invertebrates: **Freshwater**   48h EC50, *Daphnia magna* 2.0 mg/l but sensitivity depends on water hardness and temperature.

Marine   In marine systems the high salinity levels are thought to offer protection for the larvae of *Palaemonetes varians*, 0.32 mg/l (30d LC50) (Mance *et al.*, 1984c).

Algae: Concentrations of chromium VI above 1.0 mg/l may affect the growth of some algal species (Mance *et al.*, 1984c).

BIOACCUMULATION   Chromium is moderately accumulated. Aquatic algae give BCFs in the range of 3-31400 based on dry weights. Low BCFs have been recorded for fish. Laboratory studies on marine species report BCFs of 383 to 620, but levels in the field have not been found to be as high as this (WHO, 1988).

ENVIRONMENTAL FATE AND PERSISTENCE   Chromium enters aquatic systems via industrial effluent and natural processes. It is liable to precipitate out into sediments. Cr(III) and Cr(VI) are thought to be the most important oxidation states. The trivalent form of chromium forms stable complexes with neutral and negatively charged inorganic and organic species. The main removal process for chromium in waters is adsorption to suspended solids and the proportion of chromium in the water column rarely exceeds 50%. It is oxidised very slowly to CrVI (Mance *et al.*, 1984c).

STANDARDS   Chromium is a List 2 substance under the Dangerous Substance Directive (76/464/EEC). The DoE have set an EQS of 5-50ug/l (Annual Average, dissolved) in the freshwater environment and 15ug/l (Annual Average, dissolved) in the marine environment. The range for the freshwater environment reflects the hardness of the receiving watercourse as toxicity is dependent on water hardness. Standards for chromium may also be included in other legislation.

COPPER

CAS NUMBER	7440-50-8

FORMULA

Cu

USE

Copper is used in the manufacture of alloys, metal plating, wire and pipes, glass and ceramics, wood preservatives, paints, antifouling paints, as an agrochemical, and as a catalyst in vinyl chloride production. UK production was 136 000 tonnes in 1981 (Mance *et al.*, 1984a).

TOXICITY

The reported LC50 values for sensitive aquatic organisms are typically less than 100 ug/l

Invertebrates: Freshwater

48h LC50 *Daphnia magna* 0.06 mg/l (Mance *et al.*, 1984a)

ENVIRONMENTAL FATE AND PERSISTENCE

Copper exists in water either in the dissolved form as the cupric ion or complexed with inorganic anions such as carbonates and chlorides, humic and fulvic acids (Mance *et al.*, 1984a). It can also be adsorbed by particulate matter or to sediments. The concentration of each of these forms will be dependent on a number of environmental variables such as pH, salinity and hardness. At pH's in most freshwaters basic copper carbonate and cupric hydroxide would precipitate out of solution at concentrations greater than 0.5mgCu/l if there were no inorganic anions present to form complexes. Estuaries are thought to be the major depositional site for particulate copper transported by rivers (Mance *et al.*, 1984a).

STANDARDS

Copper is a List 2 substance under the Dangerous Substance Directive (76/464/EEC). The DoE have proposed an EQS of 1-28ug/l (Annual Average, dissolved) for the freshwater environment and 5ug/l (Annual average, dissolved) for the marine environment. The range for the freshwater environment reflects the hardness of the receiving water as toxicity of copper is dependant on water hardness. Standards for copper may also be included in other legislation.

1,2-DICHLOROETHANE

CAS NUMBER	107-06-2
CHEMICAL NAME	1,2-dichloroethane
SYNONYMS	EDC, 1,2-bichloroethane, ethane dichloride, ethylene chloride, ethylene dichloride, 1,2-ethylene dichloride, sym-(metric)-dichloroethane
ISOMERS	Alpha and beta 1,2-dichloroethane
FORMULA	$C_2 H_4 Cl_2$

USE

1,2-dichloroethane is a synthetic compound made from ethychlorine. In the EC production capacity was estimated at 9446 kt/y in 1983 (SRI, International, 1984). It is produced in the UK and entry into the environment is thought to predominantly occur through releases from chemical manufacturing plants (Eggersdorfer and Frische, 1983). The major use of this chemical is in the synthesis of vinyl chloride although it is also used to produce other chemicals such as 1,1,1 trichloroethane and trichloroethene. In addition it is used as a solvent for fats, oils and waxes and it is also used as a fumigant. Emissions of 1,2-dichloroethane into water amount to 0.1% of the total production (WHO, 1987).

TOXICITY

1,2-dichloroethane is of low toxicity to aquatic species.

Fish: Freshwater

Lepomis macrochirus (bluegill sunfish), 96h LC50, 430 mg/l (Buccafusco *et al.*, 1981)

Invertebrates: Freshwater

48h LC50, *Daphnia magna*, 270 mg/l (Leblanc, 1980)

BIOACCUMULATION

Bioconcentration of 1,2-dichloroethane is thought to be unlikely. A BCF of 2 was found for the *Lepomis macrochirus* (bluegill sunfish) (Verschueren, 1983; Barrows *et al.*, 1980).

ENVIRONMENTAL FATE AND PERSISTENCE

The half life in water is estimated at between 1 and 2 days with volatilisation being the major removal process. Degradation under anaerobic conditions is slow and an estimated half life of 72 years has been predicted in a closed aquatic system (Jeffers *et al.*, 1989). The reported degradation products are formyl chloride, hydrogen chloride, carbon dioxide, carbon monoxide and monochloroacetyl chloride. Slow biodegradation occurs in water (WHO, 1987).

STANDARDS

1,2-dichloroethane is a List 1 substance under the Dangerous Substances Directive (76/464/EEC). The EC have set a statutory EQS of 10ug/l for all waters. Standards for 1,2-dichloroethane may also be inlcuded in other legislation.

DICHLORVOS

CAS NUMBER	62-73-7	

Structure:

MeO — P(=O)(OMe) — O — CH === CCl$_2$

(MeO)$_2$P(=O)—O—CH=CCl$_2$

CHEMICAL NAME
Phosphoric acid 2,2-dichloro-ethenyl dimethyl ester

SYNONYMS
phosphoric acid 2,2-dichlorovinyl dimethyl ester; O,O-dimethyl O-(2,2-dichlorovinyl) phosphate; SD 1750; Astrobot; Atgard; Canogard; Dedevap; Dichlorman; Divipan; Equiguard; Equigel; Estrosol, Herkol; Norgos; Nuvan; Task; Vapona; Verdisol, Dichlorfos.

FORMULA
$C_4 H_7 Cl_2 O_4 P$

USE
Dichlorvos is a soluble organophosphorous insecticide which has uses as a fumigant in crop protection and for controlling louse in the salmon farming industry in Scotland. The latter represents the greatest use of dichlorvos in the UK. EC production was estimated at 2000-4000 t/y in 1988 (SRI, 1984).

TOXICITY
Dichlorvos is highly toxic to fish and aquatic invertebrates with 96h LC50 in the range of 0.17-11.6 mg/l for fish and 0.07-15 ug/l for invertebrates

Fish: Freshwater
48h LC50, *Lepomis macrochirus* (bluegill sunfish), 0.00007 mg/l
96h LC50, *Lepomis macrochirus* (bluegill sunfish), 0.869 mg/l (Verschueren, 1983)

Invertebrates: Freshwater
48h LC50 *Daphnia pulex*, 0.07 ug/l (WHO, 1989)

BIOACCUMULATION
Due to its high toxicity dichlorvos would kill an organism before it was taken into the tissues. Even at low concentrations which are toxic, dichlorvos seldom persists for more than one week. There is no evidence for bioaccumulation of this substance.

ENVIRONMENTAL FATE AND PERSISTENCE
Very rapid disappearance from water due to high volatility. It is detected at low concentrations in the aquatic environment. Dichlorvos is not very persistent in aqueous environments with a half life of 19-79 hours (Grectiko *et al.*, 1983). In aqueous solution dichlorvos is hydrolysed to phosphoric acid. Dichlorvos is not readily biodegraded by micro-organisms in sewage, but is found to be non-toxic to these organisms. If released on land, dichlorvos will leach into groundwater where it will hydrolyse and also degrade through chemical and biological processes with reported half-lives ranging from 1.5 to 17 days (Howard, 1990).

STANDARDS
Dichlorvos is a Red List substance and the DoE have proposed an EQS of 0.001ug/l (Annual Average) for the freshwater environment and 0.04ug/l (Annual Average) for the marine environment. Standards for dichlorvos may also be included in other legislation.

DIELDRIN

CAS NUMBER 00060-57-1

CHEMICAL NAME (1aa,2b,2aa,3b,6b,6aa, 7b,7aa)-3,4,5,6,9,9-hexachloro a,2,2a,3,6, 6a,7,7a-octahydro-2,7:3,6-dimethanonapth [2,3-6] oxinene.

SYNONYMS alvit, dieldrex, dieldrite, octaloxpanoram, quintox, kill-germ, dethlac, dieldrine, HEDD

ISOMERS Dieldrin

FORMULA $C_{12} H_8 Cl_6 O$

USE Dieldrin is an organochlorine insecticide. It is a stereo-isomer of endrin. It was previously widely used as an insecticide but was banned in the UK in May 1989.

TOXICITY Dieldrin is highly toxic to fish and invertebrates.

Fish: Freshwater LC50 96h *Pimephales promelas*, (fathead minnow), 16 ug/l (Verschueren,1987)

Invertebrates: Freshwater 48h LC50 *Daphnia pulex*, 251ug/l (Allan, 1981)

BIOACCUMULATION Readily bioconcentrated in aqueous organisms with BCFs in excess of 2000, eg. Catfish have a BCF of 3500 (Shammon, 1977). The log Kow is 4.09 (USEPA, 1992). Dieldrin is also readily taken up from food.

ENVIRONMENTAL FATE AND PERSISTENCE Dieldrin is extremely persistent in the environment and accumulates in organisms. It is very persistent in soils with a half life of 3-25 years (Verschueren, 1983). It has a half life of 723 days in river water (USEPA, 1988). Dieldrin is insoluble in water and only detected at ng/l levels in drinking water. It is formed from aldrin by metabolic oxidations in animals and chemical oxidation in soils. It has been reported in surface waters at 0.1 ug/l but rarely in ground water due to its low mobility in soils. Dieldrin does not readily undergo photolysis, and is strongly adsorbed to sediments (Boucher and Coe, 1972). The major route of removal is volatilisation (Verschueren, 1983).

STANDARDS Dieldrin is a List 1 substance under the Dangerous Substance Directive (76/464/EEC). The EC have set a statutory EQS of 10ng/l for all waters. Standards for dieldrin may also be included in other legislation.

63

DDT

CAS NUMBER	50-29-3
CHEMICAL NAME	1,1,1-trichloro-2,2-bis (4-chlorophenyl) ethane

SYNONYMS

p,p-DDT,
dichlorodiphenyltrichloroethane,
1,1-(2,2,2-trichloro-ethylidene) bis [4-chlorobenzene] dichlorodiphenyltrichloroethane;
1,1,1-trichloro-2,2-bis (p-chlorophenyl) ethane, Hildit, Digmar

ISOMERS

pp-DDT, op-DDT

FORMULA

$C_{14} H_9 Cl_5$

USE

DDT is an organochlorine insecticide. DDT has been banned in the UK since 1984 but may be present on imported goods eg. fleeces.

TOXICITY

DDT is highly toxic to fish and invertebrates.

Fish: Freshwater

96 LC50 *Pimephales promelas* (fathead minnow), 19 ug/l
96 LC50 *Lepomis macrochirus* (bluegill sunfish), 18 ug/l
96 LC50 *Oncorhynchus mykiss* (rainbow trout), 7 ug/l
15d LC50 *Oncorhynchus mykiss* (rainbow trout) 0.26ug/l
96 LC50 *Salmo trutta* (brown trout), 2 ug/l
96 LC50 *Perca fluviatilis* (perch), 9 ug/l (Verschueren, 1983)

Invertebrates: Freshwater

48h IC50, *Daphnia sp.* 4 ug/l
96h LC50, *Daphnia sp.* 1 ug/l (Sax, 1985)

BIOACCUMULATION

DDT is soluble in fats where it accumulates. BCFs of 10 000 - 40 000 are typical in fish and a BCF of 70 000 has been reported for *Daphnia* species (Verschueren, 1983; Sax, 1985). DDT is also biomagnified up food chains.

ENVIRONMENTAL FATE AND PERSISTENCE

DDT is extremely persistent. In soils DDT has a half life of 4-30 years. It is resistant to degradation by micro-organisms in activated sludge. It is biotransformed aerobically to DDE and anaerobically to DDD. It is subject to some hydrolysis and decomposition in light. DDT is insoluble in water (3.1ug/l) and found only in trace levels (0.01ug/l) in surface waters (WHO, 1989). DDT is adsorbed onto sediments. The major route of DDT removal from the aquatic environment is volatilisation (Eichelberger and Lichtenberg, 1971).

STANDARDS

DDT is a List 1 substance under the Dangerous Substance Directive (76/464/EEC). The EC have set a statutory limit for all waters of 10ng/l (para-para DDT) and 25ng/l (Total DDT). Standards for DDT may also be included in other legislation.

ENDOSULFAN

CAS NUMBER 115-29-7

CHEMICAL NAME (1,4,5,6,7,7-hexachloro-8,9,10-trinorbon-5 en-2, 3ylene dimethyl sulphite.

SYNONYMS benzoepin, thiodan, Beosit, Chlorthiepin, Cyclodon, FMCS4 62, Inseclophene, Thiofor.

FORMULA $C_9 H_6 Cl_6 O_3 S$

USE Endosulfan is a broad spectrum insecticide and acaricide. The technical product is a mixture of the two isomers alpha and beta. Annual production in the EC was less than 5000t in 1988 and EC usage of this chemical accounted for 375t/y. Less than 13t/y is used in the UK (Dequinze *et al*, 1984).

TOXICITY High acute toxicity to aquatic organisms, especially fish with typical 96h LC50 values of 1ug/l. The alpha form is more toxic than the beta form. In river water endosulfan is degraded in 4 weeks in sunlight.

Fish: Freshwater 96h LC50 *Pimephales promelas* (fathead minnow), 1.5ug/1 (WHO,1984)

Marine 96h LC50 *Ictalurus punctatus* (channel catfish), 1.5ug/l

Invertebrates: Freshwater 24h LC50, *Daphnia sp.* 240ug/l
48h LC50, *Daphnia sp.* 60ug/l
96h LC50, *Daphnia sp.* 52.9ug/l (Verschueren, 1983; WHO, 1984)

BIOACCUMULATION BCFs of 100s to 1000s have been reported. Endosulfan has a higher solubility in water than other organochlorine pesticides, but a lower affinity for fats and hence bio-accumulation of endosulfan is not likely to be as high (WHO, 1984).

ENVIRONMENTAL FATE AND PERSISTENCE Endosulfan is moderately persistent in soils and it is readily adsorbed with a low leaching potential. It is converted to endosulfan sulphate in soil. In water it has a half life of 4 days (WHO, 1984). The main degradation product is the non toxic endosulfan diol. Endosulfan is removed from water by volatilisation. Endosulfan contamination in the aquatic environment is not widespread.

STANDARDS: Endosulfan is a Red List compound and the DoE have proposed an EQS of 0.003ug/l (Annual Average) and 0.3ug/l (Maximum Allowable Concentration) for the freshwater environment and 0.003ug/l (Annual Average) for the marine environment. Standards for endosulfan may also be included in other legislation.

ENDRIN

CAS NUMBER	72-20-8
CHEMICAL NAME	1,2,3,4,10,10-hexachloro-IR,4s,4as,5s,6,7R,8R,8aR,-octahydro-6,7-epoxy-1,4:5,8-dimethanonaphthalene
SYNONYMS	endrine, endrex, ENT 17, 251, hexadrine, mendrin, nendrin
ISOMERS	Endrin
FORMULA	$C_{12} H_8 Cl_6 O$

USE

Endrin has sometimes been seen as a replacement for other more persistent organochlorine insecticides such as DDT and the 'drins'. Endrin was formerly used as an insecticide and rodenticide(USEPA, 1985). It has been banned in the UK since 1984.

TOXICITY

Endrin is highly toxic to aquatic organisms.

Fish: Freshwater

96h LC50 *Oncorhynchus mykiss* (rainbow trout), 0.6 ug/l
96h LC50 *Lepomis macrochirus* (bluegill sunfish), 0.6 ug/l
96h LC50 *Pimephales promelas* (fathead minnow), 1.0 ug/l
96h LC50 *Oncorhynchus kisutch* (coho salmon), 0.5 ug/l
(Verschueren, 1983)

Invertebrates: Freshwater

48h LC50 *Daphnia pulex,* 20 ug/l (Verschueren, 1983)

BIOACCUMULATION

Endrin is readily accumulated in fatty tissues with reported BCF values of 1335 to 10,000 in fish and 500-1250 in invertebrates. Endrin has a log Kow of 4.56 (Howard, 1990) Moderate bioconcentration has been reported in algae (Howard, 1990). Biomagnification of endrin through food chains has also been reported (Verschueren, 1983).

ENVIRONMENTAL FATE AND PERSISTENCE

Endrin is insoluble in water. Endrin is a highly persistent pesticide with a half life of 460 days in soil for aerobic degradation and 130 days for anaerobic degradation. Endrin is strongly adsorbed to soils, and is resistant to leaching. It is only detected in surface waters at low levels but persists in river sediments for a long time. A half-life of greater than 14 years for volatilisation from a model pond has been calculated (Howard, 1990).

STANDARDS

Endrin is a List 1 substance under the Dangerous Substances Directive (76/464/EEC). The EC have set a statutory standard of 5ng/l for all waters. Standards for endrin may also be included in other legislation.

FENITROTHION

CAS NUMBER 122-14-5

CHEMICAL NAME O,O-dimethyl O-4-nitro-m-tolyl thiophosphate

SYNONYMS Fenitrothion, Phosphoro-thioic acid O,O-dimethyl O-(3-methyl-4-nitrophenyl)ester; O,O-dimethyl O-4-nitro-m-tolyl phosphorothioate; MEP; Metathion, Bayer 41831; Bayer S Accothion; Cyten; Cyfen; Folithion; Sumithion

FORMULA $C_9 H_{12} NO_5 PS$

USE Fenitrothion is a broad spectrum organophosphorous insecticide. Total UK usage is estimated to be less than 5 t/y. Total EC production capacity in 1986 was 500-1000 t/y (Hedgecott, 1991). The main input route of fenitrothion into the aquatic environment is via diffuse sources. Entry via point sources may occur at manufacturing and formulation plants.

TOXICITY Invertebrates are the most sensitive species. Fish are less sensitive to fenitrothion.

Fish: Freshwater   96h LC50, *Salmo salar* (Atlantic salmon), 1.23 mg/l
96h LC50, *Oncorhynchus mykiss* (rainbow trout), 1.28 mg/l
(Verschueren, 1983)

   Marine   96h EC50, *Crassostrea virginica* (juvenile oysters), 450 ug/l
96h LC50, *Oryzias latipes* (killifish), 2.1 mg/l
(Hedgecott, 1991)

Invertebrates: Freshwater   48h EC50, *Daphnia magna* 13 ug/l (Hedgecott, 1991)

BIOACCUMULATION Fenitrothion has a moderate bioaccumulation potential. In fish BCF values range from 30 for the mullet to 540 for the killifish. Bioaccumulation in freshwater invertebrates is lower than in plants or fish. In algae BCFs are higher than 100. Bioaccumulation by marine organisms is similar to freshwater organisms (Hedgecott, 1991).

ENVIRONMENTAL FATE AND PERSISTENCE Persistence appears to be quite low with reported half lives in water ranging from 1 hour to 10 days. Complete elimination is thought to take a few weeks. Fenitrothion is readily degraded by micro-organisms in sludge, soil and water.

STANDARDS Fenitrothion is a Red List substance and the DoE have proposed an EQS for the freshwater and marine environments of 0.01ug/l (Annual average) and 0.25ug/l (Maximum Allowable Concentration). Standards for fenitrothion may also be included in other legislation.

FENTHION

CAS NUMBER	55-38-9
CHEMICAL NAME	O,O-dimethyl O-4-methyl-thio-m-toly-phosphoro-thioate

SYNONYMS	Bay 29439, Baycid, Bayer 9007, Baytex, ent 25540, entex, lebaycid, mercaptophos, MPP, OMS 2, phenthion, Queletox, S1752, Spotton, Sulfidophos, Talodex, Tiguvon
FORMULA	$C_{10} H_{15} O_3 P S_2$
USE	Fenthion is an organophosphorous insecticide with a 500-1000 t/y annual production. It is not approved for use in the UK.
TOXICITY	Fenthion is of high toxicity to aquatic organisms.
Fish: Freshwater	6h LC50 *Oncorhynchus mykiss* (rainbow trout) 0.9 mg/l (Verschueren, 1983)
Invertebrates:	Values for marine and freshwater biota are 0.0003-1.8 and 0.00002-0.5 mg/l respectively.
Freshwater	48h LC50, *Daphnia pulex*, 0.8ug/l (Verschueren, 1983)
Algae: Freshwater	96h EC50 (inhibition of growth) *Scenedesmus subspicatus* 0.05 mg/l 26% reduction in oxygen evolution, *Skeletonema costatum*, 10ppb (Verschueren, 1983)
BIOACCUMULATION	Fenthion is expected to bioaccumulate in aquatic biota.
ENVIRONMENTAL FATE AND PERSISTENCE	Fenthion is stable to hydrolysis with half lives at pH 4, pH7 and pH9 of 223 days, 200 days and 151 days respectively.
STANDARDS	No EQS has been derived within the UK. Standards for Fenthion may be included in other legislation, ie. other than the Dangerous Substance Directive (76/464/EEC).

GAMMA HCH (LINDANE)

CAS NUMBER   58-89-9

CHEMICAL NAME   1a,2a,3b,4a,5a,6b - Hexachlorocyclo-
hexane

SYNONYMS   HCH, BHC, Gamma HCH, Gamma benzene hexachloride, Benhaxachlor, Viton, Gamexane, Gexane, bbb, Ben-Hex, Aphtiria, Aprasin, Strennex, Tri-6, Lindane, Lindatox, Lorexane, Kwell, Quellada, Jacutin

FORMULA $C_6 H_6 Cl_6$

USE   Lindane is an organochlorine pesticide and wood preservative. Lindane is used as an insecticide, pediculicide (lice), scabicide (scabies) and ectoparasiticide. Point sources of contamination result from formulation plants and wood treatment installations. Diffuse sources of freshwater contamination arise from the use of lindane as an insecticide and biocide (Jones *et al*, 1988).

TOXICITY   Lindane is very toxic to aquatic organisms with LC50s typically in the low ug range.

Fish: Freshwater   96h LC50 *Oncorhynchus mykiss* (rainbow trout), 22-27 ug/l
96h LC50 *Lepomis macrochirus* (bluegill sunfish), 22 ug/l
96h LC50 *Salmo trutta* (brown trout), 68-77 ug/l
96h LC50 *Pimephales promelas* (fathead minnow), 59-87 ug/l
(Marcelle and Thome 1983; Johnson and Finley 1980; Jones *et al* 1988; Janarden *et al* 1984; Verschueren 1983; The Pesticide Manual 1987)

Invertebrates: Freshwater   24h LC50 *Daphnia. magna*, 1.15 24 ug/l (Verschueren, 1983)

Algae: Cell Multiplication Inhibition of *Microcystis aeruginosa*, occurs at 0.3 mg/l (Verschueren, 1983)

BIOACCUMULATION   Lindane is bioaccumulated in aqueous organisms with reported BCFs of 337-727 in *Cyprinodon variegatus*.

ENVIRONMENTAL FATE AND PERSISTENCE   Lindane is persistent and relatively immobile in soils and has a low affinity for water. It has an aqueous half life of 190 days. Lindane is hydrolysed and biodegraded slowly. Lindane is volatilised slowly from soil. In aqueous environments 30% of the lindane present is adsorbed to sediments (Howard, 1987).

STANDARDS   Lindane is a List 1 substance under the Dangerous Substance Directive (76/464/EEC) and the EC have set a statutory EQS of 100ng/l for the freshwater environment and 20ng/l for the marine environment. Standards for lindane may be included in other legislation.

HEXACHLOROBENZENE

CAS NUMBER 118-74-1

CHEMICAL NAME Hexachlorobenzene

SYNONYMS Perchlorobenzene, HCB, Pentachloro-phenyl chloride

FORMULA C_6Cl_6

USE Hexachlorobenzene was previously used as a fungicide but is no longer approved in the UK for agricultural use (banned in 1975) (MAFF/HSE, 1988). It is commonly used in the chemicals industry for the manufacture of aromatic fluorocarbons, chlorinated solvents and synthetic tyres.

TOXICITY Hexachlorobenzene is highly toxic to aquatic organisms with LC50 values commonly less than 1 mg/l.

Fish: Freshwater 48h LC50, *Brachydanio* sp.(zebra fish), less than 0.03mg/l (Calimari *et al.,* 1983)
14d LC50 *Lebistes reticulatus* (guppy), 0.32mg/l (Verschueren, 1983)

Invertebrates: Freshwater 24 LC50 *Daphnia* species, <0.03 mg/l (Calamari *et al.,* 1983)

Algae: 50% inhibition of photosynthesis in *Chlorella pyrenoidosa,* 5.0 mg/l (Verschueren, 1983)

BIOACCUMULATION Hexachlorobenzene is readily bioaccumulated in fish and aquatic organisms. BCFs of 5500 - 21,900 for a variety of fish have been reported.

ENVIRONMENTAL FATE AND PERSISTENCE Hexachlorobenzene is fairly persistent and resistant to biodegradation but is subject to slow photo-oxidation in freshwater (Verschueren, 1983). Hexachlorobenzene is readily adsorbed to sediments (Schaurette *et al.,* 1982).

STANDARDS Hexachlorobenzene is a List 1 substance under the Dangerous Substance Directive (76/464/EEC) and the EC have set a statutory EQS of 0.03ug/l for all waters. Standards for hexachlorobenzene may be included in other legislation.

HEXACHLOROBUTADIENE

CAS NUMBER 87-68-3

CHEMICAL NAME 1,1,2,3,4,4-Hexachloro-1-3-butadiene

SYNONYMS Hexachloro-1,3-butadiene, perch-lorobutadiene, HCBD;1,3-hexachlorobutadiene

FORMULA $C_4 Cl_6$

USE Hexachlorobutadiene is a by-product in the production of tetra-chloroethylene, trichloroethylene, carbon tetrachloride and chlorine. It is also used as an intermediate in the production of lubricants and rubber compounds (IARC, 1979). EU production was 2000-4000 t/y in 1988.

TOXICITY Acute toxicity to marine organisms occurs at concentrations as low as 32ug/l.

Fish: Toxicity of hexachlorobutadiene to fish is high with LC50 values less than 1mg/l.

Freshwater 14d LC50, *Lebistes reticulatus* (guppy), 0.5 mg/l (Verschueren, 1983)

BIOACCUMULATION BCF values are in the range of 100-1000. There is no evidence that hexachlorobutadiene accumulates through food chains.

ENVIRONMENTAL FATE AND PERSISTENCE Hexachlorobutadiene has a high stability and is inert at ambient temps. Physical and chemical degradation occur in water but are slow. It readily becomes adsorbed to sediments. Hexachlorobutadiene is insoluble in water, and is expected to readily adsorb to sediments in aquatic systems. The half life in surface water is reported to range from 4 weeks to 6 months (Howard, 1991).

STANDARDS Hexachlorobutadiene is a List 1 substance under the Dangerous Substance Directive (76/464/EEC) and the EC have set a statutory EQS of 0.1ug/l for all waters. Standards for hexachlorobutadiene may also be included in other legislation.

71

LEAD

CAS NUMBER

7439-92-1

FORMULA

Pb

USE

Lead occurs chiefly as a sulphide in galena and is used in the manufacture of batteries, cable cover, solder, pigments, building material, radiation shields and cement (WHO, 1977).

TOXICITY

The toxicity of lead varies depending on availability, uptake, and species sensitivity.

Fish: Freshwater

Oncorhynchus mykiss (rainbow trout) 96h LC50 1.32 mg/l (Davies *et al.,* 1976)

Invertebrates: Freshwater

48h LC50, *Daphnia magna,* 0.45 mg/l (lead chloride) (Calabrese *et al.,* 1973; Biesinger and Christensen, 1972)

      Marine

48h LC50, *Crassostrea virginica* (eastern oyster), 2.45 mg/l

Algae:

No effect observed at concentrations below 1-15 mg/l

BIOACCUMULATION

There is no evidence to suggest that lead is biomagnified through food chains. BCFs of up to 100 000 have been reported (WHO, 1989).

ENVIRONMENTAL FATE AND PERSISTENCE

Inputs of lead are from diffuse sources. Lead is insoluble and its availability is limited by its strong adsorption to sediments and organic matter. The uptake of lead is slow and is influenced by environmental factors. Environmental contamination by lead is widespread (WHO, 1989).

STANDARD

Lead is a List 2 compound under the Dangerous Substance Directive (76/464/EEC). The DoE have proposed a standard of 4-20ug/l (Annual Average, Dissolved) in the freshwater environment and 25ug/l (Annual Average, Dissolved) for the marine environment. The range for the freshwater environment reflects the fact that the toxicity of lead is dependent on water hardness. Standards for lead may also be included in other legislation.

MALATHION

CAS NUMBER 121-75-5

CHEMICAL NAME S-1,2-bis(ethoxycarbonyl ethyl-0,0-dimethyl phosphorodithioate)

SYNONYMS diethyl [(dimethoxyphosphinothioyl) thio]succinate, Carbofos, Maldiron, Mercaptothion

FORMULA $C_{10} H_{19} O_6 P S_2$

USE Malathion is an organophosphorous insecticide used for broad spectrum control in agriculture and forestry (FAO/WHO). Malathion is not manufactured in the UK.

TOXICITY Some commercial formulations are reported to be more toxic than the active ingredient alone. Malathion is highly toxic to fish with LC50 values in the range of 0.1-10 mg/l and 0.76ug/l - 3mg/l for invertebrates.

Fish: Freshwater 96h LC50 *Oncorhynchus kisutch* (coho salmon), 0.10 mg/l
96h LC50 *Lepomis macrochirus* (bluegill sunfish), 0.10 mg/l
96h LC50 *Pimephales promelas* (fathead minnow), 8.7 mg/l (Verschueren, 1983)

Invertebrates: Freshwater 96h LC50, *Gammarus fasciatus* 0.76 ug/l-1
48h LC50, *Daphnia pulex* 1.8 ug/l-1 (Sanders and Cope, 1966; Sanders, 1972)

BIOACCUMULATION Malathion is rapidly metabolised by fish and is not readily bioaccumulated, with a BCF of 2 for the freshwater fish *Pseudorasbora parva*.

ENVIRONMENTAL FATE AND PERSISTENCE Malathion is not very persistent. It has a half life in soil of 3-7 days and is rapidly degraded in aqueous environments (Howard, *et al.,* 1991). It is degraded by bacteria under both aerobic and anaerobic conditions. Malathion may enter the aquatic environment via spray drifts, aerial applications (although not in the UK), run-off or discharge from industrial and sewage effluent. It is not readily leached from soils due to rapid microbial degradation.

STANDARDS Malathion is a Red List substance. The DoE have proposed an EQS of0.01ug/l (Annual Average) and 0.5ug/l (Maximum Allowable Concentration) in the freshwater environment and 0.02ug/l (Annual Average) and 0.5ug/l (Maximum Allowable Concentration) in the marine environment. Standards for malathion may also be included in other legislation.

MERCURY

CAS NUMBER	7439-97-6
FORMULA	Hg
USE	Its major use is as a cathode in electrolysis of sodium chloride solution to produce caustic soda and chlorine gas. It is also used in measuring instruments, pharmaceuticals, dentistry, catalysts, cathodes and antifouling paints. Different species of mercury have varying properties but methyl mercury is the most hazardous (WHO, 1976).

TOXICITY

Fish:	Freshwater	*Lebistes reticulatus* (guppy) acute toxicity threshold of 30 ug/l
	Marine	*Platichthys flesus* (flounder) acute toxicity threshold 1.5 ug/l
Invertebrates:	Freshwater	*Daphnia* species acute toxicity 2 ug/l Shrimp acute toxicity 3.5 ug/l

BIOACCUMULATION

The BCFs for mercury II and methyl mercury are 4994 and 4000-85 000 respectively. The BCF for oysters in the marine environment range from 10 000-40 000 for both mercuric chloride and methyl mercury.

ENVIRONMENTAL FATE AND PERSISTENCE

In aquatic environments organomercury compounds and mercurous oxide appear to be 4 to 30 times more toxic than mercury(II). Data suggests that mercury is converted to methyl mercury by micro-organisms. Elimination of methyl mercury from fish and from other aquatic organisms is slow. Loss of inorganic mercury is more rapid. Sea birds can accumulate mercury from contaminated food supplies. Mercury accumulates in sediments where it is converted to methyl mercury by bacteria. Methyl mercury can enter food chains and accumulates in top predators (WHO, 1976).

STANDARD

Mercury is a List 1 substance under the Dangerous Substance Directive (76/464/EEC). The EC have set a statutory EQS of 1ug/l (Total) for the freshwater environment and 0.3ug/l (Dissolved) for the marine environment. Standards for mercury may also be included in other legislation.

NICKEL

CAS NUMBER	7440-02-0
FORMULA	Ni
USE	Nickel is used in the production of alloys, electroplating, as a pigment in ceramics and in the production of azodyes; it is also a catalyst. Production of nickel is 14600 t/y in the EC (Mance and Yates, 1984).

TOXICITY

Fish: Freshwater — 96h LC50, *Oncorhynchus mykiss* (rainbow trout), 7 mg/l (Mance and Yates, 1984)

Marine — 96h LC50 for the *Menidia menidia* (Atlantic silverside), 8 mg/l (WHO, 1991).

Invertebrates: Acute EC50s and LC50s range from 0.64-4.96mg/l (EPA, 1980)

Marine — The lowest adverse effect on marine crustaceans were observed for *Mysidopsis bahia* where the 96h LC50 was reported to be 0.51 mg/l (Mance and Yates, 1984)

Algae: Freshwater algae do not appear to be sensitive to nickel. The lowest reports are a 96h EC50 of 0.36mg/l for *Scenedesmus subspicatus*. (WHO, 1991)

BIOACCUMULATION

Nickel is not bioaccumulated to any significant extent in freshwater fish. For algae BCFs range from 0.9-808. Invertebrate species appear to accumulate higher concentrations of nickel than fish. For *Daphnia magna* BCFs up to 157 have been recorded. Bioconcentration levels in marine organisms appears to be low except for molluscan species where BCFs of 1300-2080 have been recorded (WHO, 1991).

ENVIRONMENTAL FATE AND PERSISTENCE

Entry into aquatic systems is via diffuse and point sources. Nickel adsorbs to organic matter and sediments. It is liable to leach from soils. Nickel is highly insoluble in water but can complex with a variety of ligands to form soluble complexes. Nickel shows a lower tendency than most metals to become associated with solid phases. In marine waters it is predominantly removed by adsorption to sediment particles (Mance and Yates, 1984).

STANDARDS

Nickel is a List 2 substance under the Dangerous Substance Directive (76/464/EEC). The DoE have proposed an EQS of 50-200ug/l (Annual Average, Dissolved) for the freshwater environment and 30ug/l (Annual average, Dissolved) in the marine environment. The range in the freshwater environment reflects the fact that the toxicity of nickel is dependent on the water hardness. Standards for nickel may be included in other legislation.

NITRATE

FORMULA

NO_3^-

USE

Nitrate is a negative ion, not a discrete substance. It is combined with a positive ion, usually sodium (Na^+) or potassium (K^+). Nitrates are present naturally in soils, waters and all plant materials. The use of nitrogenous fertilisers and intensive farming practices increase these natural levels. Nitrates are also used in the production and preservation of cured meat and fish products. Concentrations of nitrates in water depends on geochemical conditions, human and animal waste management practices, the use of nitrogen fertilisers and industrial discharge of nitrogen compounds. Petroleum refineries, and food and fuel processing industries are thought to constitute and important source of nitrates.

ENVIRONMENTAL FATE AND PERSISTENCE

There are a number of pathways by which nitrogen can be interchanged between the atmospheric and terrestrial forms, this is known as the nitrogen cycle. Factors affecting this cycle are climatic conditions, type and density of animal and plant populations, agricultural practices, and animal husbandry. Plants assimilate only part of the nitrates present in soil. Leaching will occur into ground waters and rivers. Also denitrification will convert nitrates to nitrogen or nitrous oxides (WHO, 1976). Although nitrates are present in surface waters higher concentrations are thought to occur in ground waters.

STANDARD

No EQS has been derived within the UK. Standards for nitrate may be included in other legislation, ie. other than that relating to the Dangerous Substance Directive (76/464/EEC).

PARATHION

CAS NUMBER   56-38-2

CHEMICAL NAME   O,O- diethyl O-4- nitro-
                      phenyl phosphorothioate

SYNONYMS   Diethyl 4-nitrophenyl
                      phosphorothionate

FORMULA $C_{10}H_{14}NO_5PS$

USE   Parathion is a synthetic broad spectrum insecticide. It has not been used in the UK.

TOXICITY

Fish: Freshwater   96h LC50, *Pimephales promelas* (fathead minnow),1.41mg/l (Priester, 1965)

Invertebrates: Freshwater   48h LC50, *Daphnia pulex,* 0.60ug/l (Verschueren, 1984)

BIOACCUMULATION   Bioconcentration of parathion in aquatic organisms is expected to be low to moderate due to its removal to sediments. The following BCF values apply to parathion; *Lepomis macrochirus* (bluegill sunfish) 81,187, 253 and 27 after 12h, 29h, 46h, and 504 days. In *Fundulus heteroclitus* (mummichog) a BCF of 80 has been reported (Verschueren, 1983).

ENVIRONMENTAL FATE AND PERSISTENCE   Parathion adsorbs strongly to soil particles and decays by biological and chemical hydrolysis over a period of several weeks. In water parathion becomes bound to sediment where hydrolysis can occur (Howard, 1990).

STANDARD   No EQS has been derived within the UK. Standards for nitrate may be included in other legislation, ie. other than that relating to the Dangerous Substance Directive (76/464/EEC).

PARATHION-METHYL

CAS NUMBER 298-00-0

CHEMICAL NAME O,O- dimethyl O-4- nitro-phenyl phosphorothioate

MeO — P( )O — [benzene ring] — NO$_2$ with S double bonded, MeO groups on P

SYNONYMS Parathion-methyl; phos-phorothioic acid O,O-dimethyl O-(4-nitrophenyl)ester; O,O dimethy O,O-p nitrophenyl thiophosphate; dimethyl parathion; metaphos; E 601; ENT-17292; Folidol-M; Metacide; Metron; Nitrox 80; Penncap M.

FORMULA $C_8 H_{10} NO_5 P S$

USE Parathion-methyl is an insecticide. Annual EC production is 6000 t/y. It has not been used in the UK.

TOXICITY

Fish: Freshwater 96h LC50, *Oncorhynchus mykiss* (rainbow trout), 2.8 mg/l (Palawski *et al*, 1983)

Invertebrates: Freshwater 48h EC50, *Daphnia magna*, 7.8-9.1ug/l (Dortland, 1980)

BIOACCUMULATION Parathion-methyl does not readily bioconcentrate and is rapidly metabolised.

ENVIRONMENTAL FATE AND PERSISTENCE Parathion-methyl is moderately soluble and moderately adsorbed onto soil particles. It should degrade by photolysis and biodegradation in soil over the course of several months. Aquatic degradation of methyl parathion takes between 2 and 4 weeks.

STANDARD No EQS has been derived within the UK. Standards for parathion-methyl may be included in legislation other than that relating to the Dangerous Substance Directive. (76/464/EEC).

PENTACHLOROPHENOL

CAS NUMBER	87-86-5
CHEMICAL NAME	Pentachlorophenol
SYNONYMS	Chlorophen, PCP, penchlorol, penta, pentachlorofenol, pentachlorofenolo, pentachlorophenate
ISOMERS	2,3,4,5,6-pentachlorophenol, pentanol
FORMULA	$C_6 Cl_5 OH$
USE	Pentachlorophenol is extensively used as a wood preservative. EC production is estimated at 108 000t/y.
TOXICITY	Pentachlorophenol is of high toxicity to aquatic organisms with LC50s of less than 1 mg/l for fish

Fish:	Freshwater	*Oncorhynchus mykiss* (rainbow trout) 24h EC50 0.115 mg/l (Mayer and Ellersieck, 1986)
Invertebrates:	Freshwater	24h LC50 *Daphnia magna* 0.92 mg/l (Mayer and Ellersieck, 1986)
Algae:	Freshwater	Toxic to *Chlorella pyrenoidosa* at 0.001 mg/l. 96h LC50, *Selenastrum capricornutum*, 0.5 mg/l (Verschueren, 1983)

BIOACCUMULATION	A BCF of 475 has been reported for *Carassius auratus* (goldfish) (Verschueren, 1983).
ENVIRONMENTAL FATE AND PERSISTENCE	Pentachlorophenol dissociates in water at ambient pHs. The dissociated form will photodegrade (with a half life in the range of hours to days). Biodegradation probably becomes significant after a period of acclimatisation. Adsorption to sediments is reported to be considerable. Hydrolysis and volatilisation are not thought to be important processes in water (Howard, 1990).
STANDARD	Pentachlorophenol is a List 1 substance under the Dangerous Substance Directive (76/464/EEC). The EC have set a statutory EQS of 2ug/l in all waters. Standards for pentachlorophenol may also be included in other legislation.

PCBs

CAS NUMBER

CHEMICAL NAME   Polychlorinated biphenyls

SYNONYMS   Chlorinated biphenyls, Arochlor, Clophen, Fenclor, Kanechlor, Phenochlor, Pyralene, Santotherm: these are all trade names containing mixtures of various PCB's (Chloride groups will replace the hydrogens in the biphenyl diagram to create a PCB).

FORMULA $C_{12} H_{(10-n)}-Cl_n$

PRODUCTION   PCBs are identified by a 4 digit code, the last two digits indicating the degree of chlorination. EU production is estimated to have been at 2000-4000t/y, but production has now ceased. They are used as heat transfer fluids, fire resistant hydraulic fluids and the dielectric medium in transformers. Approximately 50% of the PCBs produced enter the environment through disposal of industrial effluent into rivers and coastal waters and through runoff from landfills.

TOXICITY   **Fish:**   Acute freshwater toxicity occurs at levels exceeding 3ug/l and at around 10ug/l in marine waters. The effect of PCBs in fish is cumulative and the toxicity appears to decrease with increasing chlorination.

Freshwater   *Oncorhynchus mykiss* (rainbow trout), 5d LC50, 156 ug/l

Invertebrates: **Freshwater**   *Daphnia magna,* toxicity threshold for survival is 5 ug/l

BIOACCUMULATION   The following BCFs have been recorded; pink shrimp 6600, oyster 8100, *Lepomis macrochirus* (bluegill sunfish) 71400. Highest tissue levels have been found in top predators.

ENVIRONMENTAL FATE AND PERSISTENCE   PCBs are extremely stable in the environment although a small proportion are broken down by photolysis. PCBs are very persistent and strongly absorbed onto sediments. PCBs are absorbed by mammals and are subject to biomagnification through food chains. Most PCBs entering the aquatic environment are thought to be retained by particulate matter.

STANDARD   No EQS has been derived for PCBs within the UK. Standards for PCBs may be included in legislation other than that relating to the Dangerous Substance Directive (76/464/EEC).

SIMAZINE

CAS NUMBER 122-34-9

CHEMICAL NAME 6-chloro-N2,N4-diethyl-1,3,5-triazine-2,4-diamine

SYNONYMS AT,2-chloro-4,6-bis (ethylamino)-5-triazine,6-chloro-N,N-diethyl-1,3,5-triazine-2,4-diamine, 2-chloro-4,6-bis(ethylamino)-1,3,5-triazine, besaton, primatols, aquazine, simane, simadex

FORMULA $C_7 H_{12} Cl N_5$

USE Simazine is a systemic triazine herbicide which is used against annual weeds and is available in a wide variety of formulations. All triazines used in the UK are imported but there are several UK formulation and distribution sites. UK usage is estimated at between 50-200t/y in agriculture and 3000t/y in total vegetative control (MAFF/HSE, 1988). It has been banned for non-agricultural use since September 1993.

TOXICITY

Fish: The results for fish toxicity are very variable; in general simazine is considered to be of slight to moderate toxicity to fish.

Freshwater 96h LC50 *Lepomis macrochirus* (bluegill sunfish), 90mg/l
96h LC50 *Oncorhynchus mykiss* (rainbow trout), >100mg/l
48h LC50, *Oncorhynchus mykiss* (rainbow trout), 5-56 mg/l
24h LC50, *Oncorhynchus mykiss* (rainbow trout), 5mg/l
96h LC50 *Lebistes reticulatus* (guppy), 49mg/l (The agrochemicals Handbook, 1988)

Invertebrates: Freshwater 48h EC50, *Daphnia sp.* 1-92 mg/l
48h LC50, *Daphnia* species1 mg/l (Sanders, 1970; Verschueren, 1983)

BIOACCUMULATION Simazine has a low tendency to bioaccumulate and has a recorded BCF value of 2 for *Lepomis macrochirus* (bluegill sunfish).

ENVIRONMENTAL FATE AND PERSISTENCE Simazine is fairly persistent in water and soils and is slow to degrade in soil. It has a half life of between 3 weeks and 3 months in soil and one month in water. Simazine is biodegraded slowly by micro-organisms to produce hydroxy derivatives which are less toxic and subject to further degradation. It is slowly hydrolysed in water and is also subject to photolysis. Simazine is found in both aquatic and terrestrial compartments. Simazine shows a high affinity for soils and the major route of degradation here is through abiotic hydrolysis. As volatilisation is minimal the main routes of removal in aquatic systems are photolysis, adsorption to sediments and degradation by micro-organisms (Hedgecott, 1989).

STANDARD Simazine is a Red List compound and the DoE have proposed an EQS of 2ug/l (Annual Average) and 10ug/l (Maximum Allowable Concentration) for both the freshwater and marine. Standards for simazine may also be included in other legislation.

TETRACHLOROETHYLENE

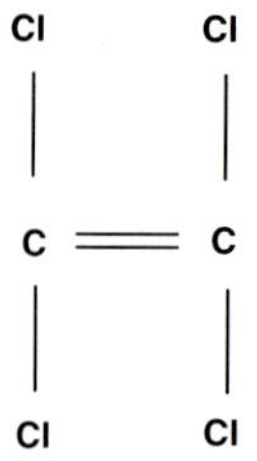

CAS NUMBER 127-18-4

CHEMICAL NAME Tetrachloroethene

SYNONYMS carbon-bichloride, carbon dichloride, 1,1,2,2-tetra-chloroethylene; ethylene tetrachloride, Perca fluviatilis (perch)loroethylene

FORMULA C_2Cl_4

USE Tetrachloroethylene is widely used as a solvent in dry cleaning and in the manufacture of paint remover. It is a synthetic chemical with numerous industrial releases into the environment. It may also be formed during water chlorination.

TOXICITY Tetrachloroethylene is moderately toxic to aquatic organisms

Fish: **Freshwater** 96h LC50 *Oncorhynchus mykiss* (rainbow trout), 4.99 mg/l
96h LC50 *Pimephales promelas* (fathead minnow), 18.4mg/l (Alexander *et al.*, 1987)

Invertebrates: **Freshwater** 48h LC50 *Daphnia magna*, 18 mg/l (WHO, 1984a)

ENVIRONMENTAL FATE AND PERSISTENCE Tetrachlorethylene is chemically stable in aquatic solutions with a half life of six months, at 1mg/l in sunlight. At low concentrations tetrachloroethylene is slowly degraded under anaerobic conditions. The major removal process from water bodies is expected to be volatilisation (WHO, 1984a).

STANDARD Tetrachloroethylene is a List 1 substance under the Dangerous Substance Directive (76/464/EEC). The EC has set a statutory EQS of 10ug/l for all waters.

1,2,3-TRICHLOROBENZENE

CAS NUMBER	87-61-6
CHEMICAL NAME	1,2,3-trichlorobenzene
SYNONYMS	VIC-trichlorobenzene; UN2321
FORMULA	$C_6H_3Cl_3$

USE

1,2,3-trichlorobenzene is widely used as a solvent in industry and also used as a lubricant, coolant, in electrical installations and glass tempering (Verschueren, 1983).

TOXICITY

This chemical has a moderate to high toxicity to freshwater species

Fish: Freshwater   48h LC50 *Oncorhynchus mykiss* (rainbow trout), 0.71 mg/l

Invertebrates: Freshwater   24h EC50, *Daphnia magna*, 0.35 mg/l (Calamari *et al.*, 1983)

Algae: Freshwater   EC50 (growth inhibition) *Selenastrum capricornutum*, 0.9 mg/l

BIOACCUMULATION

1,2,3-trichlorobenzene is lipophillic and is readily bioconcentrated in fish with a BCF value of 13000 for the *Lebistes reticulatus* (guppy) after 19 days exposure to 48ug/l (Verschueren, 1983).

ENVIRONMENTAL FATE AND PERSISTENCE

1,2,3-trichlorobenzene is readily degraded by bacteria. It is readily removed from the aquatic environment by volatilisation. 1,2,3-trichlorobenzene is present in the environment as a consequence of industrial processes but can also be formed as breakdown product of other chlorinated organics.

STANDARD

Trichlorobenzene is a List 1 substance under the Dangerous Substance Directive (76/464/EEC). The EC have set a statutory EQS of 0.4ug/l for all waters. Standards for trichlorobenzene may also be included in other legislation.

1,1,1-TRICHLOROETHANE

CAS NUMBER   71-55-6

CHEMICAL NAME   1,1,1-trichloroethane

SYNONYMS   Aerothene TT, chlorothene NU, chlorten, chlorothane NU, chlorothene, VG, ox-trichloroethane (T), chloroethene, chlorotene, chlorothene, methylchloroform, methyltrichloro-methane, inhibisol, solvent 111 extreme grade

FORMULA $C_2H_3Cl_3$

USE   1,1,1-trichloroethane is used as an industrial solvent, and as an industrial cleaning agent. It is also an intermediate in vinylidene production. It is a synthetic chemical with no natural sources.

TOXICITY **Fish:** 1,1,1-trichloroethane has a low toxicity to fish with typical LC50 values of more than 50 mg/l

Freshwater   96h LC50 *Pimephales promelas* (fathead minnow), 53.4 mg/l (Verschueren, 1983)

Invertebrates: **Freshwater**   IC50 *Daphnia magna* 37.6 mg/l (Deneer *et al*, 1988)

BIOACCUMULATION   1,1,1-trichloroethane is not expected to be bioaccumulated to a significant extent (Howard, 1990).

ENVIRONMENTAL FATE AND PERSISTENCE   1,1,1-trichloroethane inhibits anaerobic degradation at low concentrations. It has a half life, in surface waters, in the range of hours to weeks depending on the wind and mixing conditions. It takes several weeks to degrade in soils. 1,1,1-trichloroethane occurs as a pollutant in air, water and foodstuffs. It volatilises readily from water.

STANDARD   Trichloroethane is a List 1 substance under the Dangerous Substance Directive (76/464/EEC). The DoE have proposed an EQS of 0.1mg/l (Annual Average) and 1.0mg/l (Maximum Allowable Concentration) for the freshwater environment. No standard has been proposed for the marine environment.

TRICHLOROETHYLENE

CAS NUMBER	79-01-6	

CHEMICAL NAME   Trichloroethene

SYNONYMS   Acetylene trichloride, ethylene trichloride, TCE, ethinyl trichloride;

FORMULA $C_2 H Cl_3$

USE   Trichloroethylene is a commonly used solvent in industry it is also used in anaesthetics. It is a synthetic chemical with no known natural sources.

TOXICITY   **Fish:**   Trichloroethylene shows a low acute toxicity to fish

Freshwater   96h LC50, *Pimephales promelas* (fathead minnow), 40.7 mg/l

Invertebrate: **Freshwater**   40h LC50, *Daphnia sp.* 600 mg/l

BIOACCUMULATION   Trichlorethylene is not bioaccumulated in freshwater organisms to any great extent and in marine organisms it is only moderately bioconcentrated. For *Lepomis macrochirus* (bluegill sunfish) a BCF of 17-39 has been recorded (Howard, 1990).

ENVIRONMENTAL FATE AND PERSISTENCE   Trichloroethylene is not very persistent with an aqueous half life in the range of minutes to hours with most of the removal via volatilisation. It is adsorbed to sediments. Trichloroethylene has a wide distribution in the environment and is found in water, air and food. It may be introduced into surface and groundwaters by industrial effluents but is generally released into the atmosphere, it volatilises readily from surface waters (Howard, 1990).

STANDARD   Trichloroethylene is a List 1 substance under the Dangerous Substance Directive (76/464/EEC). The EC have set a statutory EQS 10ug/l for all waters. Standards for trichloroethylene may also be included in other legislation.

TRIFLURALIN

CAS NUMBER 1582-09-8

CHEMICAL NAME trifluoro-2,6-dinitro-N,N-dipropyl-p-toluidine

SYNONYMS 2,6-dinitro-N,N-dipropyl-4-triflouro-methylaniline; 2,6-dinitro-N,-N-dipropy -4-(trifluoro-methyl) benzamine (CA); trifluraline

FORMULA $C_{13} H_{16} F_3 N_3 O_4$

USE Trifluralin is a dinitroaniline herbicide used for the control of broad leaved weeds. EC production is estimated at 1000-3000 t/y of which 6t/y is used in the UK

TOXICITY

Fish: Trifluralin is highly toxic to fish with LC50 values typically less than 1mg/l

Freshwater 96 LC50 *Onchorhynchus mykiss* (rainbow trout), 10-40 ug/l
96 LC50 *Lepomis macrochirus* (bluegill sunfish), 20-90ug/l

Marine 96 LC50 *Ictalurus punctatus* (channel catfish), 45-440ug/l (Verschueren, 1983; Manual of Acute Toxicity, 1986)

Invertebrates: Freshwater 96h LC50 *Daphnia magna*, 1 - 2.8 mg/l-1
48h LC50 *Daphnia pulex*, 0.24mg/l (Verschueren, 1983)

Algae: High acute toxicity for algae. 97% inhibition of growth in *Oedogonium cardiacum* over a 30 day exposure period at 0.022 mg/l (Yockim *et al.*, 1983).

BIOACCUMULATION Trifluralin gives rise to a BCF of 1800-6000 in fish (Verschueren, 1983).

ENVIRONMENTAL FATE AND PERSISTENCE Trifluralin is moderately persistent with a half life of 3-57 days in water. Trifluralin loss from the water phase through volatilisation will be slow. Trifluralin has a low water solubility of 0.05-24 mg/l and high affinity for soil particles; hence groundwater contamination is rare as it is resistant to leaching. Biodegradation and photo degradation may produce polar metabolites which may contaminate drinking water. Maximum concentrations reported in sea water and groundwater are 16 ug/l and 0.54ug/l respectively. Trifluralin is very stable in water and likely to be adsorbed to sediments due to low solubility. Probable pollution sources are from production and formulation plants, spillage during transport, handling and storage and spray drift, run-off and farm waste (Jones, 1990).

STANDARD Trifluralin is on the UK Red List and the DoE have proposed an EQS of 0.1ug/l (Annual Average) and 20ug/l (Maximum Allowable Concentration) in both the freshwater and marine waters. Standards for trifluralin may also be included in other legislation.

TRIORGANOTIN COMPOUNDS

The organotins of environmental concern are tetrabutyltins, tributyltins (TBT) and triphenyltins (TPT). TBTs are mostly used in antifouling paints but also as wood preservatives. TPTs are mostly used as agricultural fungicides but also in antifouling paints. Tetrabutyltin is used as an intermediate for the production of tri, di and monobutyltin.

In aqueous environments they tend to come out of solution and bind to particulate matter and sediments. In particular they can become associated with phytoplankton. Subsequent biomagnification can then occur, (Anon, 1985).

TRIPHENYLTINS

FORMULA

Triphenyltinchloride $C_{18} H_{15} ClSn$ (639-58-7)
Triphenyltinacetate $C_{20} H_{18} O_2 Sn$ (900-95-8)
Triphenyltinhydroxide $C_{18} H_{16} OSn$
Triphenyltinfluoride $C_{18} H_{15} FSn$ (379-52-2)

```
          Ph
          |
Ph ——— Sn ——— []
          |
          Ph
```

USE

TPT production in the U.K. is estimated to be 70 t/y (1988). TPT itself is an industrial intermediate used to form TPT acetate, hydroxide, chloride and fluoride. Possible pollution sources arise at the production and formulation plants, during transport and handling. Coastal and diffuse inputs arise mostly from diffuse sources.

TOXICITY

Fish: Freshwater
96h LC50, *Cyprinus carpio*, (carp), 19 ug/l (TPT-acetate)
96h LC50, *Oncorynchus mykiss*, (rainbow trout), 22 ug/l (TPT-hydroxide)

Invertebrate: Freshwater
48h LC50, *Daphnia magna*, 10 ug/l (TPT-hydroxide)

BIOACCUMULATION

BCFs of 300-400 have been recorded

ENVIRONMENTAL FATE AND PERSISTENCE

Half-life under field conditions is 6-19 days. Degradation is mainly by photolysis and biodegradation, (Waldock, *et al.*, 1988). In the environment triphenyltins readily hydrolyse to a triphenyltin-cationic complex which is the same for all triphenyltin compounds.

STANDARD

Triphenyltin compounds are List 2 substances under the Dangerous Substance Directive (76/464/EEC) and the DoE have proposed EQS of 0.02ug/l (Total) for the freshwater environment and 0.008ug/l (Total) for the marine environment. These standards are expressed as Maximum Allowable Concentrations. Standards for triphenyltin may also be included in other legislation.

FORMULA

$C_{12} H_{27} Sn$

$$
\begin{array}{c}
\text{Bu} \\
| \\
\text{Bu} \longrightarrow \text{Sn} \longrightarrow [\,] \\
| \\
\text{Bu}
\end{array}
$$

USE

TBT production in the U.K. is estimated to be 800 t/y (1988). Used in anti-fouling paints although their use for this application is now decreasing due to restricitons on its use.

TOXICITY

For fish 96h LC50 values range from 1.5-35 ug/l. The toxicity increases with the number of butyl groups (Laughlin *et al.*, 1986).

Fish: Freshwater

168h LC50, *Lebistes reticulatus* (guppy), 13-240 ug/l

Invertebrates: Marine

TBTs are highly toxic to marine molluscs causing shell deposition in oysters and affecting gonad development in gastropods (imposex). *Eurytemorra affinis* (copepods) are more sensitive than other crustaceans displaying 96h LC50 values of 0.6-2.2 ug/l (Bryan *et al.*, 1986).

Freshwater

48h LC50, *Daphnia sp.* 2.3-70 ug/l

BIOACCUMULATION

BCFs of up to 7000 have been recorded in molluscs and fish. The primary route of uptake is food. BCFs for molluscan species range from 2000-11000 (Wood, 1986).

ENVIRONMENTAL FATE AND PERSISTENCE

TBTs are found mostly in estuaries resulting from diffuse sources. TBT adsorbs readily onto particles. Physical, chemical and biological degradation occurs but is dependant on temperature, pH and oxygen levels. Dibutyl derivatives are formed which are more readily degraded than TBT. In water TBT dissociates to become a positively charged anion. The inorganic radical has little effect on biocidal activity of the organotin compound. TBT concentrates in the surface microlayer where concentrations are 2-27 times higher than in the subsurface layers (Davies *et al.*, 1987). TBT absorbs onto particles and is subject to transformation resulting from physico-chemical and biochemical processes. TBT is chemically degraded by progressive debutylation, hydrolysis and photodegadation. It is also biodegraded. TBT is slow to degrade with a half life of 1-3 weeks in aerobic conditions and several years in anaerobic conditions. BCF values for TBT range from 1000 to >10 000 for macro-organisms and > 30 000 in algae and bacteria. TBT accumulates in sewage sludge, sediments and biota. Concentrations in the marine environment are typically 1-100 ng/l but may be >350 ng/l in marinas. Freshwater concentrations of >70 ng/l have been reported.

STANDARD

Tributyltin compounds are List 2 compounds under the Dangerous Substance Directive (76/464/EEC) and the DoE have proposed EQS of 0.02ug/l (Total) in the freshwater environment and 0.002ug/l (Total) in the marine environment. These standards are expressed as maximum allowable concentrations. Standards for tributyltin may also be included in other legislation.

ZINC

CAS NUMBER 7440-66-6

FORMULA Zn

USE Zinc is used in the manufacture of alloys, electroplating, textiles, paper production, fungicides, ceramics, glass and in paints. UK production in 1981 was 82 000 tonnes (Mance and Yates, 1984a)

TOXICITY

Fish: Freshwater 96h LC50, *Fundulus heteroclitus* (mummichog), 60 000ug/l (Eisler, 1967)
48h LC50 *Salmo salar* (Atlantic salmon) smolt, 35 000ug/l (Herbert and Wakeford, 1964)

Invertebrate: Freshwater 48h LC50 *Daphnia magna* 0.16 mg/l (Biesinger and Christensen, 1961)

BIOACCUMULATION Biomagnification factor of three for the mussel, *Mytilis edulis.* BCF values for marine molluscs range from 670-16700 (Mance and Yates, 1984a).

ENVIRONMENTAL FATE AND PERSISTENCE Zinc forms complexes with ammonia, amines, halides, cyanides, and other inorganic and organic ligands. Zinc is one of the most ubiquitous and mobile of heavy metals and is transported in natural water systems and to a great extent in rivers in the dissolved form. Adsorption to sediments and particulate matter is dependant on pH, alkalinity and salinity.

STANDARD Zinc is a List 2 substance under the Dangerous Substance Directive (76/464/EEC) and the DoE have proposed EQSs for both the freshwater and marine environment. The proposed standard for the freshwater environment is 8-125ug/l (Annual average, total) with the range reflecting the fact that the toxicity of zinc is dependent on water hardness. For the marine environment an EQS of 40ug/l (Annual average, dissolved) has been derived for the marine environment. Standards for zinc may also be included in other legislation.

REFERENCES

Alexander, H. C., McCarty, W. M. and Bartlett, E. A. (1978). Toxicity of perchloroethylene, trichloroethylene, 1,1,1-trichloroethane and methylene chloride to fathead minnows. Bulletin of Environmental Contamination and Toxicology, 20, (3), 344-352.

Anon (1985). TBT support document. Office of Pesticides Programmes, ESEPA, Washington DC.

Barrows, M. E., Petrocelli, S. R., and Caroll, J. T. (1980). Bioconcentration and elimination of selected water pollution by sunfish *(Lepomis macrochirus)*. In: Dynamics, exposure and hazard assessment of toxic chemicals. Ann Arbor Science. Ann Arbor, MI. (Cited in Waldbridge *et al.,* 1983).

Biesenger, K. E. and Christensen, G.M. (1972). Effects of various metals on the survival growth, reproduction and metabolism of *Daphnia magna*. Journal of Fisheries Research Board Canada, 29, 1691-1700.

Boucher, F. R. and Coe, G. F. (1972). Adsorption of lindane and dieldrin pesticides on unconsolidated aquifer sands. Environmental Science and Technology, 6, 60, 538-543.

Bryan, G. W., Gibbs, D. E., Hummerstone, L. G. and Burt, G. R. (1986). The decline of the gastropod *Nucella lapillus* around the South West of England: evidence of the effect of TBT from antifouling paints. Journal of the Marine Biological Association, 66, 611-640.

Buccafusco, R. J., Ellis, S. J., Leblanc, G. A. (1981). Acute toxicology of priority pollutants to the *Lepomis macrochirus* (bluegill sunfish). Bulletin of Environmental Contamination, 26, (4), 446-452.

Butijn, G. D., Koeman, J.H. (1977). Evaluation of possible impact of aldrin, dieldrin and endrin on the aquatic environment. Dept. Toxicology, Agricultural University, Wageningen, The Netherlands.

Calabrese, H., Collier, R. S., Nelson, D. A. and McInnes, J. R. (1973). The toxicity of heavy metals to embryos of the American oyster *Crassostrea virginica*. Marine Biology, 18, 162-166.

Calamari, D., Galsi, S, Selt, F, and Vighi, M. (1983). Toxicity of selected chlorobenzenes to aquatic organisms. Chemosphere, 12, 253-266.

Davies, I. M. (1987). TBT in Scottish sea lochs, as indicated by the degree of imposex in the dog whelk, *Nucella lapillus.* Marine Pollution Bulletin, 18, (7), 400-404.

Deneer, J. W., Sinnige, T. L., Seinen, W., Hermans, J. L. M. (1988). The toxicity of aquatic pollutants: QSARS and mixture toxicity to *Daphnia magna* of industrial chemicals at low concentrations pp 81-87.

Department of the Environment (DoE) (1992). River Quality The Government's Proposals: A Consultation Paper. DoE, December 1992.

Department of the Environment (DoE) (1990). United Kingdom North Sea Action Plan 1985 - 1995. Presented to the Third North Sea Conference held in the Hague, March 1990. Department of the Environment.

Department of the Environment (DoE) (1988). Ministerial Declaration. 2nd International Conference on the Protection of the North Sea, London 24/25 November 1987.

Department of the Environment (DoE) (1989). Agreed "Red List" of dangerous substances confirmed by the minister of State (Lord Caithness). News Release 194, DoE, Marsham Street, London, 10 April 1989.

Dequinze, J., Scimar, C. and Edeline, F., (1984). Identification of the substances and their derived products on the list of 129 substances (List 1 of the Directive 76/464/EEC), present in the refuse of chlorine derived organic chemical industry. Final Rep. to the Community of the EC XI15/85, 221-222.

Dilling, W. L. (1975). Evaporation rates and reactivities of dichloromethane, chloroform, 1,1,1-trichloroethane, trichloroethylene and tetrachloroethylene. Current Research, 9, (9), 933.

Dortland, R. J. (1980). Toxicological evaluation of parathion and azinphos methyl in freshwater model ecosystems. In. Versl. Landbouwk. Onderz, 898, 112p.

Eggersdorfer, R., and Frische, R. (1983). Study of the discharges of certain chloro- and bromoethanes into the aquatic environment and the best technical means for the reduction of water pollution from such discharges. Final report for the Commission of European Communities. Contract no. U/82/177 (540). Batelle Institute, Frankfurt, Oct. 1983, CEC XI/813/83.

Eichelberger, J. W. and Lichenberg, D. (1971). Persistence of pesticides in river water. Environmental Science and Technology, 5, (6), 541-544.

EPA, (1990). Ambient water quality criteria for nickel. EPA 440/5-80-060. USEPA.

Ferguson, J. F. and Gavis, J. (1972). A review of the arsenic cycle in natural waters. Water Research, 6, 1259.

Grectiko, A. V., Nabolotnaya, O. M. and Marchenko, P. V. (1983). The hydrolysis of dichlorvos. Soviet Journal of water Chemistry and Technology, 5, (2), 87-89.

Hartman, W. A., Martin, D. B. (1985). Effect of four agricultural pesticides on Daphnia pulex, Leuna minor, Potomogeton pectinatus. Bulletin of Environmental Contamination, 35, 646-651.

Hedgecott, S. (1989). Proposed provisional environmental quality standards for atrazine and simazine in water (ESSL 9378 SLG) DoE 2149 WRc Report.

Hedgecott, S. (1991). Proposed provisional environmental quality standards for fenitrthion in water DoE 2197 WRc Report.

Helling, C. S., Zhangw, B. A., Gish, T. J.,Coffman, C. B., Isensee, A. R., Kearney, P. C., Hoagland, D. R., Woodward, M. D. (1988). Persistence of atrazine, alachlor and cyanzine under no tillage practices. Chemosphere, 17, (1), 175-187).

Henriet, J., Detroux, L. and Agie de Selasten, J. (1989). Study of the technical and economical aspects of measures to reduce water pollution caused by the discharge of atrazine, bentazone and chloroprene. Final Report to CEC, 16 Oct, 1987.

Howard, P. H. (1990). Fate and exposure data for organic chemicals: Pesticides. Lewis Publ.

Howard, P. H. (1990). Fate and exposure data for organic chemicals: Solvents. Lewis Publ.

Howard, P. H. (1990). Fate and exposure data for organic chemicals: Priority pollutants. Lewis Publ.

Howard, P. H. and Boethling, R. S. (1991). Handbook of environmental degradation rates. Lewis Publ., Michigan, USA.

Janarden, S. K., Oslan, C. S. and Schaerren, D. J. (1984). Quantitative comparisons of acute toxicity of organic chemicals to rats and fish. Ecotoxicology and Environmental Safety, 8, 531-539.

Jeffers, P.M., Ward, L. M., and Wolfe, N. L. (1989). Homogenous hydrological rate constants for selected chlorinated methanes, ethanes, ethenes and propanes. Environmental Science and Technology, 23, 965-969.

Johnson, W. W. and Finely, M. T. (1980). Handbook of acute toxicity of chemicals to fish and aquatic invertebrates. US dept. of the Interior, Fish. and W·.dl. Serv. Resource Publ 137.

Jones, A., Hedgecott, S. and Zabel, T. (1988). Red list report. WRc report PRU-190/m2

Jones, A. (1990). Proposed environmental quality standards for trifluralin in water. WRc Report DoE 2231-m/1.

Laughlin, R. B., (1986). Bioaccumulation of TBT: the link between the environment and organisms. Oceans 86. Conference Record 4, organotin symposiun 1206-1209.

Leblanc, G. A. (1980). Acute toxicology of priority pollutants to th water flea. Bulletin of Environmental Contamination, 24, (5), 684-689.

Mailhot, H, Peters, R.H. (1988). Empirical relationship between the octanol/water partition coefficient and 9 physicochemical properties. Environmental Science and Technology, 22, 1479-1488.

MacKay, D., Paterson, S., Cheung, B. and Neely, W.B. (1985). Evaluation of the environmental behaviour of chemicals with a level III fugacity model. Chemosphere, 14, 335-374.

Mance, G., Brown, V. M. and Yates, J. (1984). Proposed environmental quality standards for list II substances in water: copper. WRc report TR210

Mance, G., Brown, V. M., Gardiner, J. and Yates, J. (1984c). Proposed environmental quality standards for list I substances in water: chromium. WRc Technical Report, TR207.

Mance, G. and Yates, J. (1984). Proposed environmental quality standards for list II substances: Nickel. WRc Technical Report, TR 211.

Marcelle, C. and Thome, J. P. (1983). Acute toxicity and bioaccumulation of lindane in gudgen. Bulletin of Environmental Contamination and Toxicology, 31, 453-458.

Manual of Acute Toxicity (1986). Interpretation and data base for 410 chemicals and 66 species of freshwater animals. US Dept of the Interior, Resource publ. 160, Washington DC.

MAFF/HSE (1988) Pesticides 1988. Pesticides approved under the Control of Pesticides Regulations 1986.

Mayer, F. L. and Ellersieck, M. R. (1986). Manual of acute toxicity; interpretation and database for 410 chemicals and 66 species of freshwater animals. US Dept. Interior, Fish and Wildlife Service, resource Public. No. 160.

Moorhead, D. L., Koswinski, R.J. (1986). Effect of atrazine on the production of artificial stream algal communities. Bulletin of Environmental Contamination, 37, 330-336.

NRA (1991a). NRA Baseline Estuary and Coastal Waters Monitoring Programme. National Rivers Authority, Anglian Region, 2 January 1991.

NRA (1991b). The Paris Commission 1991. Survey: Report for England and Wales. National Rivers Authority, Anglian Region.

NRA (1990). The Paris Commission 1990. Survey: Report for England and Wales. National Rivers Authority, Anglian Region.

Palawski D., Buckler, D. R. and Mayer, F.L (1983). Survival and condition of rainbow trout *(Salmo gairdneri)* after acute exposures to methyl-parathion, triphenyl phosphate and DEF. Bulletin of Environmental Contamination and Toxicology, 30 (5), 614-620. PARCOM (1992). Monitoring and Assessment. Oslo and Paris Commissions. July 1992.

PARCOM (1988). Principles of the Comprehensive Study on Riverine Inputs. Convention for the Prevention of Marine Pollution from Land Based Sources. Tenth Meeting of the Paris Commission, Lisbon 15 - 17 June 1988.

Penrose, W.R. (1974). Arsenic in the marine and aquatic environments.; analysis, occurrence and significance. Critical Reviews on Environmental Control, Oct 1974.

Priester, L. E. (1965). The accumulation and metabolism of DDT, parathion and endrin by aquatic food chain organisms. Zoology, 1661-B.

PSL (Produce Studies Limited) (1990). The use of herbicides tin non-agricultural situations in England and Wales, Rep. No. PSL. 5813/DJC/DWP, Sept 1990.

Sanders, H. O. (1972). Toxicities of some herbicides to six species of freshwater crustaceans. Journal of Water Pollution Control Federation, 42, 1544-1550.

Sanders, J.G. (1979). The concentration of arsenic in marine macro-algae. Estuarine and Coastal Marine Science, 9, 95-.

Sanders and Cope (1966). Toxicities of several pesticides to two species of cladocerans. Trans Amer. Fish Soc. 92, 165.

Sax, I. (1985). Dangerous properties of industrial materials, (5), (1), 12-20. New York, 1985.

Schaurette, W., Lay, L. P., Kleine, W. and Korte, F. (1982). Long term, fate of organochlorine xenobiotics in aquatic ecosystems. Ecotoxicology and Environmental Safety, 6, 560-569.

SDIA, (1989). Detergent phosphate and water quality in the UK. Booklet produced by the Soap and Detergent Industry Association, PO Box 9, Hayes Gate House, Hayes, Middlesex, UB4 OJD.

Shammon, L. R. (1977). Accumulation and elimination of dieldrin in muscle tissue of the *Ictalurus punctatus* (channel catfish). Bulletin of Environmental Contamination and Toxicology, 17 (6), 637-644.

SRI International, (1984). Directory of chemical producers. Western Europe. Volume 2. 7th ed. Merlo Park, Ca, SRI, .

The Pesticide Manual, (1987). The British Crop Protection Committee. 8th ed.

USEPA, (1980). Ambient water quality criteria for ars   Ref No. PB82-117327

USEPA, (1988). Ambient water quality for dieldrin.

USEPA, (1992). Risk reduction. (PREL). Treatability database, Cincinnati, Ohio, USA.

USEPA, (1985). Health advisory on endrin, office of drinking   ter.

Verschueren, K. (1983). Handbook of environmental data on organic chemicals. 2nd ed. Van Nostrand Reinhold Public.

Walbridge, C. T., Fianatt, J. T, Phipp, G. H., and Holcombe, G. W. (1983). Acute toxicology of 10 chlorinated aliphatic hydrocarbons to the *Pinnephales promelas* (fathead minnow). Archives of Environmental Contamination and Toxicology, 12, 661-666.

Waldcock, M. J., White, M. E, and Thain, J. E. (1988). Inputs of TBT into the marine environment from shipping activity in the UK. Environmental Technology Letters, 9, 999-1010.

WHO, (1984). Guidelines for drinking water and health, 191-194.

WHO (IPCS), (1989). Environmental Health Criteria 1991, WHO Geneva.

WHO, (1976-1992). Environmental Health Criteria 1, 2, 3, 5, 18, 31, 40, 50, 61, 62, 71, 79, 83, 85, 86, 91, 108, 116, 134, 135, 136.

Worthing, C. R. (1991). The pesticide manual - a world comp   ium, 9th ed. British Crop Protection Council.

Yockim, R. S., Isensee, A. R. and Walker, E. A. (1980). Behaviour of trifluralin in aquatic model systems. Bulletin of Environmental Contamination and Toxicology, 24, 134-141.